Climate Change and Political Strategy

Although the science of climate change is well-established and there are well-known policy instruments that could significantly reduce greenhouse gas (GHG) emissions without prohibitive economic costs, political obstacles to more determined action remain despite heightened concern among mainstream politicians and the public. This book analyses the political dynamics of climate policy in affluent democracies from a number of different theoretical angles in order to improve our understanding of which political strategies would be likely to enable national governments to make deep cuts in GHG emissions while avoiding significant political damage.

This book was previously published as a special issue of *Environmental Politics*.

Hugh Compston is Reader in Politics in the School of European Studies at Cardiff University, UK. Major publications include *Policy Networks and Policy Change* (Palgrave, 2009), *King Trends and the Future of Public Policy* (Palgrave, 2006), *Handbook of Public Policy in Europe: Britain, France and Germany* (edited, Palgrave, 2004), *Social Partnership in the European Union* (edited with Justin Greenwood, Palgrave, 2001), *Policy Concertation and Social Partnership* (edited with Stefan Berger, Berghahn, 2002), and *The New Politics of Unemployment* (edited, Routledge, 1996) as well as numerous journal articles on public policy and political economy.

Climate Change
and Political Strategy

Edited by Hugh Compston

R Routledge
Taylor & Francis Group

LONDON AND NEW YORK

First published 2010
by Routledge
2 Park Square, Milton Park, Abingdon, Oxon, OX14 4RN

Simultaneously published in the USA and Canada
by Routledge
711 Third Avenue, New York, NY 10017

Routledge is an imprint of the Taylor & Francis Group, an informa business

First issued in paperback 2011

© 2010 Taylor & Francis

Typeset in Times by Value Chain, India

British Library Cataloguing in Publication Data
A catalogue record for this book is available from the British Library

ISBN13: 978-0-415-45870-2 (hbk)
ISBN13: 978-0-415-50939-8 (pbk)

CONTENTS

Notes on Contributors

Ian Bartle is the author of numerous publications in the areas of energy and telecommunications policy, risk and regulation. His most recent publication is 'Risk and the Regulatory State: A Better Regulation Perspective' (with Peter Vass), *Centre for the Study of Regulated Industries Report No. 20* (University of Bath, 2008).

Catherine Butler is a Research Associate at the School of Psychology, Cardiff University, UK. Her publications include 'Public Understanding of Climate Change and the Media' (with Nick Pidgeon), in T. Boyce and J. Lewis (eds.), *Climate Change and the Media* (Peter Lang, 2009), and 'Risk and the Future: Floods in a Changing Climate,' *21st Century Society* 3(2) (2009).

Hugh Compston is a Reader in Politics at the Cardiff School of European Studies, Cardiff University, UK. Publications include *Policy Networks and Policy Change* (Palgrave, 2009), *Turning Down the Heat: The Politics of Climate Policy in Affluent Democracies* (ed. with Ian Bailey) (Palgrave, 2008), and *King Trends and the Future of Public Policy* (Palgrave, 2006).

Amy Fletcher is a Senior Lecturer in Political Science at the University of Canterbury, New Zealand. Her publications include 'Mendel's Ark: Conservation Genetics and the Future of Extinction', *Review of Policy Research* 25(6) (2008), and 'Governing DNA', in H. Gottweis and A. Petersen (eds.), *Biobanks: Governance in Comparative Perspective* (Routledge, 2008).

Neil Gavin is a Senior Lecturer in the School of Politics and Communication Studies, University of Liverpool, UK. His publications include *Press and Television in British Politics* (Palgrave, 2007), and contributions to T. Boyce (ed.), *Climate Change and the Media* (Peter Lang, 2009), and A. Charles (ed.), *Media in the Enlarged Europe* (Intellect, 2009).

Frank Grundig is a Lecturer in International Relations in the School of Politics and International Relations at the University of Kent, UK, and has published widely on the logic of international climate change negotiations.

Nick Pidgeon is Professor and Director of the Understanding Risk Research Group in the School of Psychology at Cardiff University, UK. Recent publications include *The Social Amplification of Risk* (edited with Roger Kasperson and Paul Slovic) (Cambridge, 2003) and *The GM Debate: Risk Politics and Public Engagement* (with 6 others) (2007, Routledge).

Sarah Pralle is an Associate Professor in the Department of Political Science at the Maxwell School at Syracuse University, USA. Her publications include *Branching Out, Digging In: Environmental Advocacy and Agenda-setting* (Georgetown University Press, 2006) and "'I'm Changing the Climate, Ask Me How!": The Politics of the Anti-SUV Campaign', *Political Science Quarterly* 121(3) (2006).

Ivan Scrase is a Research Fellow at SPRU - Science and Technology Policy Research, University of Sussex, UK. Publications include *Energy for the Future - A New Agenda* (ed. with G. MacKerron) (Palgrave, 2009) and 'CCS in the UK: Squaring Coal Use with Climate Change?' in J. Meadowcroft and O. Langhelle (eds.), *Caching the Carbon* (Edward Elgar, 2009).

Adrian Smith is a Senior Research Fellow at SPRU - Science and Technology Policy Research, University of Sussex, UK, and has published widely on socio-technical transitions management.

Introduction: political strategies for climate policy

Hugh Compston

This volume is based on the premise that the principal obstacles to stronger action on climate change are political in nature. The science of climate change is well-established (IPCC 2007a) and there are well-known policy instruments that could significantly reduce greenhouse gas emissions without prohibitive economic costs (Stern 2007), yet governments and other political authorities are reluctant to take decisive action even though most appear to be convinced that strong measures are needed. At present the main political strategy seems to be the implementation of measures that target a broad range of emissions sources while not antagonising business groups or electorates. Typical policies include setting emissions targets, encouraging promising technologies, using market mechanisms such as taxes and emissions trading to spur innovation, and urging greater international cooperation on climate policy. So far, however, such measures have failed to reverse the steady rise in atmospheric concentrations of greenhouse gases (GHGs) such as carbon dioxide (IPCC 2007b, p. 4, Tans 2009).

Clearly more needs to be done. But how can governments move beyond existing policies without risking serious political damage? The aim of this volume is to contribute to answering this question by analysing the nature of climate politics from a number of different theoretical angles in order to improve our understanding of which political strategies would be likely to help national governments to make deep cuts in GHG emissions while avoiding significant political damage.

The rationale for this multi-theoretical approach is that different conceptual and logical schemas (theories) highlight different features of situations. Thus, describing the politics of climate policy in terms of different theories results in different conceptual and logical pictures of this phenomenon. This means that at least to some extent the inferences drawn from these pictures about the nature of the political obstacles to more vigorous action on

Table 1.   Political strategies to enable governments in affluent democracies to take more effective action against climate change while avoiding significant political damage.

Type of political strategy	Specific strategy
Refinement of current strategies	Further efforts to reach global agreement
	Better reporting of climate change and clearer communication of the policy instruments that are needed
	Stricter emission and policy targets
	Identification and introduction of more policies on which all powerful actors can agree
	Incremental strengthening of existing policies
	Preparation of measures that can be implemented swiftly in response to public concern following extreme weather events
	Continued emphasis on the contribution of climate policies to other policy objectives, such as energy security
	More vigorous use of existing policy instruments, especially economic instruments and financial incentives to promote technological innovations and renewable energy production; more stringent voluntary agreements; and extension of emissions trading
Exploration of new policies	More stringent energy efficiency regulation and increased financial incentives for energy efficiency improvement
	Grand projet-style state investment in new infrastructure
	Personal carbon allowances
	Carbon import tariffs
Governance reform	Improved measurement of emissions
	More systematic envisioning of what a low carbon society would look like
	Integration of economic and environmental governance
	Provision of seats for independent experts and environmental NGOs on all climate-policy-related committees on which industry is represented
	Placement of able and committed individuals in key posts
	Improvements in the transparency of potentially popular initiatives
	More equitable distribution of costs
Greater emphasis on spillover policies	Policies that are easily transferable to other countries, difficult to reverse once introduced, and/or create pressure for their own strengthening or the introduction of related measures
Selective imposition of more radical policies	Introducing strong policies early in each electoral term
	Targeting economic sectors that can pass on extra costs to consumers
	Targeting losses on small sections of society
	Compensating powerful actors

Source: Compston and Bailey 2008b.

climate change, and the best ways of overcoming them, will also be different. Together, therefore, these analyses reveal a more detailed and nuanced view of the political options open to activist governments than can be obtained from studies that stick to a single theoretical perspective.

Although there are a small number of previous studies that analyse the politics of climate change using explicit theoretical perspectives (see, for example, O'Riordan and Jaeger 1996, Newell and Paterson 1998), this is the first multi-theoretical study in this area. Most academic writing on climate politics in affluent democracies (see, for example, Helm 2005, Bailey 2007) does not directly address the question of which political strategies are likely to be of greatest assistance to governments, although a number of strands touch upon it. Accounts of the politics of climate change in various Western countries often have implications for this question (see, for example, Bailey and Rupp 2005, Oshitani 2006, Kerr 2007), and studies of discourses within environmental politics draw conclusions about political communications strategies (Hajer and Versteeg 2005, Ereaut and Segnit 2006). The growing literature on climate policies at sub-national level also includes relevant observations (Rabe 2004), while the extensive literature on international climate politics often touches upon domestic factors that influence governments' negotiating positions (Dolzak 2001).

Recently, however, two major studies have appeared to which issues of political strategy are more central. The first is an empirical study of the history and politics of climate policy in affluent democracies that focused directly on this question (Compston and Bailey 2008a). The conclusions of this study are set out in Table 1.

The second is a broader study of the politics of climate change in general by Anthony Giddens (2009). His book is very much based on the premise that it is essential to get the public on board, and business too, if the required policies are to be put in place, although he does acknowledge that business interests that oppose action may need to be faced down from time to time. Along with prescriptions about what needs to be done, which among other things stress multilateral action, an active state that plans and ensures outcomes rather than merely facilitating action, technological innovation, carbon taxes, embedding concern for climate change in the daily lives of citizens, and using public–private partnerships as a way of obtaining needed finance, a number of strategies designed to surmount or circumvent political obstacles are identified. These include:

- employing information strategies that focus on a few key indicators, especially those linked to focusing events such as weather-related natural disasters;
- favouring economic and technological innovations that generate competitive advantage and thereby attract business support (economic convergence);
- preferring policies that not only combat climate change but also help to achieve other policy goals, especially energy security (political convergence);

- promoting a positive model of a low carbon future rather than one based on giving things up, for example by providing incentives for low carbon practices wherever possible rather than imposing punishments for high carbon activities;
- spreading a concern with climate change through all branches of government; and
- seeking cross-party support to ensure continuity of climate policy.

However, neither of the volumes just mentioned brings the big guns of contemporary theories of political causation to bear on this issue. The aim of this volume is to remedy that deficiency by applying a number of such theories to the specific task of identifying effective political strategies for national governments that wish to make deeper cuts in greenhouse gas emissions. The focus is mainly on the national level because that is where substantive policy measures are mostly formulated and implemented, and because the international dimension of the politics of climate change is already well covered in the international relations literature.

Each contribution to this volume describes a different theory of political causation, provides a diagnosis of the politics of climate change from the perspective of this theory, and identifies, on the basis of this diagnosis, relevant political strategies.

The first part of the volume consists of studies that stress the role of policymaking techniques.

Pidgeon and Butler address the use of risk management techniques to analyse and justify particular climate policies, as exemplified in the Fourth IPCC Assessment Report (IPCC 2007a) and the Stern Review (Stern 2007). After a wide-ranging analysis of the strengths and limitations of risk analysis as an analytic technique, the authors conclude that its widespread use can be ascribed as much to its fit with contemporary governance narratives and practices as to its appropriateness to the problem at hand, which in some critical respects is quite limited. If policymakers are to better understand and deal with the high levels of complexity and uncertainty that characterise our picture of climate change, the tools of risk analysis need to be adapted and supplemented by alternative approaches. For example, policymaking processes could be reformed to include values and rationalities other than economic ones by means of enabling a wider range of stakeholders to participate in the policymaking process – although the extent to which this is likely to happen will be limited by the fact that alternative approaches do not fit as well with prevailing governance practices as does conventional risk analysis.

Ian Bartle's contribution is aimed at showing how more careful selection of policy instruments, along with reform of the policymaking process, could lead to better results in terms of controlling greenhouse gas emissions. After an analysis of the strengths and weaknesses of major theories of regulation, in particular public interest theory, private interest theory and regime theory, Bartle considers the possibility that greater transparency in regulation could lead to more effective climate policies by means such as increasing legitimacy and facilitating policy learning, and he raises the possibility of implementing a political strategy based on this by centring regulation of emissions around a transparent carbon price instrument such as a carbon tax or a cap and trade system. The problem he identifies with this approach, however, is that market instruments appeal to just one type of human rationality, namely that of an economic actor who responds only in a self-interested way to price signals, whereas in fact there is considerable evidence that individuals and organisations use other rationalities as well. Egalitarians, for example, want greater equity between humans and between humanity and nature, while hierarchicalists want better governance and planning to ensure that the natural world and its resources are better managed. This suggests that a combination of policy instruments needs to be put in place in order to secure wide support, in particular:

- transparent market instruments, such as carbon taxes and cap and trade schemes, to appeal to *homo economicus*;
- command and control regulation, to appeal to hierarchicalists;
- information and education, to appeal to egalitarians.

While there is a danger of policy proliferation, confusion and contradiction in this approach, it is argued that it is possible for policies to be complementary and mutually reinforcing.

Scrase and Smith analyse the dynamics of climate policy using a theory specifically designed to describe and facilitate technological change in its social and economic context. The basic idea of this transitions management approach is that new socio-technical regimes emerge through successful application and learning in niche uses or 'protected spaces' in which new technologies and social practices are not exposed to the full selective pressures operating in the incumbent regime. Applying this approach to climate policy would involve beginning by setting a low carbon goal such as a national or sector emissions reduction target. A series of multi-participant 'transition arenas' would then be convened to identify those regimes where emissions are most significant or the potential for change is greatest. Each transition arena would then go through an iterative process of (1) understanding the carbon reduction challenge for the existing regime and identifying 'transition goals', such as making electricity supply more sustainable; (2) developing a consensus about alternatives and a basket of 'visions' compatible with the transition goals; (3) identifying 'pathways' towards those visions, such as expansion of renewable energy; (4) instigating niche experiments that contribute to the realisation of these pathways; and (5) establishing processes for social learning and reflexivity across all of these activities. The idea is that over time this process would lead to a gradual transfer of institutional support away from the existing regime and towards a low carbon regime. Although the authors note that this approach implies a somewhat unrealistic model of politics in which niche lessons are expected to be taken up and acted upon consensually, they do identify a number of political strategies that are consistent with it, in particular:

- investing more in research and appraisal processes;
- taking steps to elevate the standing of interactive, expert-led decision-making;
- making wider use of high-level expert assessments to justify policy;
- including policies in election manifestos in order to give them political legitimacy in the event of victory;
- creating large, powerful and well funded institutions with a remit to pursue the project's aims, while curtailing the power of institutions that may be resistant to radical change, such as government departments that are close to fossil energy companies;
- taking steps to tie future governments into continuing the political project to achieve transitions, as the British government, for instance, has already done by legislating to commit future governments to legally binding cuts in greenhouse gas emissions over the period up to 2050.

The authors note that its reliance on industry and experts suggests that the transitions management process is more closely aligned with top-down technocratic forms of corporatism than with bottom-up radical decentralisation. The significance of this observation is that it implies that low carbon transition pathways are more likely to proceed by enabling investment in large scale energy supply technologies, such as nuclear power and carbon capture and storage, than by expanding decentralised energy supply and accelerating efficiency improvements in homes.

The second part of the volume comprises two studies that take a more political approach.

The theory of political causation used in my own contribution is one of the dominant theories of policymaking today: policy network theory. The particular version used is based on the idea that networks of political actors in any given area of public policy are created and sustained by interdependencies, and that consequently public policy is largely the result of elected and appointed officials pursuing their own policy preferences while at the same time exchanging policy amendments for resources that they need but which are controlled by other actors, in particular formal approval, political support, cooperation with implementation, and private investment. The logic of this theory suggests that governments that want to strengthen climate policy have four main strategic options, each of which is accompanied by distinctive types of political tactics.

The first option is to impose climate policies without exchanging resources at all. This means no policy concessions, but is politically risky. However, this risk can be minimised by tactics such as:

- introducing unpopular policies early in their terms of office;
- imposing measures on particular industries while leaving other industries alone;
- targeting business sectors that can pass on any additional costs;
- targeting those social groups that are least able to retaliate via the ballot box.

The second strategic option is the more usual one of exchanging resources within existing parameters. Among possible tactics here are:

- putting together package deals that include concessions in other areas of public policy as well as, or instead of, amendments to the climate policy under consideration;
- taking advantage of windows of opportunity created by weather-related natural disasters that are linked by media coverage to climate change.

In some circumstances it may be possible to alter the policy preferences of other actors by means of communications strategies such as:

- providing clear and accurate information about climate change and effective responses to it;
- stressing how climate policies serve other objectives as well, such as energy security;
- adopting discourses that frame the issue of climate change in new ways, such as conceptualising the effort to control climate change, in the context of economic crisis, as a Green New Deal.

The fourth strategic option is to try to change the terms of resource exchange by means such as:

- altering the distribution of formal decision-making power within government, for example by moving responsibility for energy policy from an economic ministry to an environment ministry;
- giving new actors access to the policymaking process in the hope that this may make them more likely to accept proposed policy changes;
- acquiring new legal powers in order to offer new benefits to other actors, and thus reduce the policy concessions necessary to obtain their agreement, for example by appropriating additional planning powers for central government, or by forming new agencies to make investments in areas that private firms will not touch;

- seeking cross-party agreement in order to deny opponents of these policies the opportunity of inflicting political damage on activist governments by voting for parties that oppose these policies;
- seeking new sources of political support, help with policy implementation, and business investment.

The rational choice approach employed by Frank Grundig focuses on interactions among political actors, and seeks to explain political outcomes on the basis of a small number of assumptions and deductive reasoning. Rational choice models assume that actors are self-interested and that they maximise their utility in accordance with their preferences. Grundig applies several such models to analysing first the formation of states' domestically derived negotiation positions on climate change, and second the dynamics of these international negotiations. As a result, several promising political strategies are identified:

- taking steps to enlarge the membership of environmental non-governmental organisations in order to boost their influence on national negotiating positions;
- persuading environmental organisations to direct more funding into campaign contributions in states hesitant to take action;
- using organisations such as the World Bank to assist developing countries to strengthen civil society, on the basis that this would result in greater pressure on their governments to act, as well as allowing emissions trading to take place more widely;
- stepping up information campaigns in order to counter the influence of campaign advertising and thus reduce the influence of special interest groups opposed to action on climate change;
- re-balancing abatement costs between the EU and the USA, on the basis that previous studies indicate that the net economic benefits under the Kyoto Protocol would have been negative for the USA and positive for the EU;
- making abatement more efficient by introducing an international emissions cap and trade scheme.

The final part of the volume focuses on communications.

Neil Gavin's contribution illustrates how examination of media coverage can illuminate the politics of climate policy by analysing the role of the media in the dynamics of the public sphere, by which is meant that area of public life where common concerns and societal problems are defined and debated and, as a result, public opinion is formed. Since climate change is a global phenomenon, ultimately it is the international public sphere that is relevant here. A major theme is that the relatively low priority given by the media to coverage of climate change makes it difficult for activist governments to convince citizens that radical and costly action is needed, or for concerted pressure to be applied to governments to take such action, although coverage of IPCC reports and weather-related natural disasters is likely to bring climate change up the agenda from time to time. For this reason governments need to:

- tread warily and with circumspection;
- distance themselves from responsibility for the negative consequences of climate change intervention;
- keep direct costs to government as low as possible, lest budgets become over-stretched and other pressing concerns have to give;
- avoid direct criticism for obvious failures of omission or commission in policy implementation or integration, especially in relation to taxation.

Sarah Pralle looks at the politics of climate change through the lens of agenda-setting theories. Although elected and appointed officials may believe in climate change and want to do something about it, the fact that they are also confronted with policy problems in other areas means that whether climate change is actually high on their agenda for decision is another matter. For this reason it is vital to identify factors that will help the climate change issue rise and stay high on the agendas of governmental and non-governmental institutions. Pralle's contribution shows how this can be done by using theories of agenda-setting to identify strategies for defining the problem of climate change in ways that raise its salience with the public, on the basis that this will put governments under greater pressure to address it. These include:

- regularly reporting key problem indicators in user-friendly terms;
- emphasising scientific consensus and knowledge;
- emphasising growing public concern;
- emphasising specific local impacts and personal experience;
- emphasising human health impacts;
- inserting a moral and ethical perspective into the debate.

She also points out strategies for framing solutions in ways that attract maximum support and protect against factors that might otherwise cause the public and policymakers to abandon efforts to solve the problem, such as cynicism and fatigue. These include:

- pointing to existing solutions;
- framing solutions in terms of energy;
- emphasising the costs of doing nothing;
- focusing on economic gains associated with green technology;
- providing regular feedback about policies and progress.

Strategies for maintaining political will include:

- taking advantage of focusing events such as weather-related natural disasters;
- offering 'predigested' policies to overcome gridlock;
- venue-shopping.

In the final contribution, Amy Fletcher applies frame analysis, an approach derived from discourse theory, to identify and analyse argumentative strategies utilised by political actors in their efforts to build a consensus around action on climate change, using the USA under the second Bush Presidency as a case study. Two strategies in particular are foregrounded:

- framing climate change as a security issue in order to enlist the support of those concerned with national security;
- using the Apollo metaphor to liken the task of controlling climate change to the effort during the 1960s to put a man on the moon.

Fletcher concludes that the Apollo framing is especially promising because its positive framing of climate policy in terms of technological achievement, industrial transformation and economic opportunity appeals to a particularly wide range of political actors and voters.

More and more attention is being paid to climate change and considerable thought has now been devoted to working out how in principle it can be brought under control. Progress in understanding the politics of climate change has lagged behind somewhat, but is just as important if the required policies are to be put in place. By bringing a number of theories of political causation to bear on this problem the contributors to this volume are able jointly to generate, or at least make explicit, more ideas about strategies for activist governments than a single theoretical perspective could yield.

To understand what these ideas mean and evaluate their chances of success if implemented, however, you will need to begin by evaluating their logic within the theoretical contexts in which they have been formulated.

References

Bailey, I., 2007. Climate policy implementation: geographical perspectives. *Area*, 39, 415–417.

Bailey, I. and Rupp, S., 2005. Geography and climate policy: a comparative assessment of new environmental policy instruments in the UK and Germany. *Geoforum*, 36, 387–401.

Compston, H. and Bailey, I., 2008a. *Turning down the heat: the politics of climate policy in affluent democracies*. Basingstoke: Palgrave Macmillan.

Compston, H. and Bailey, I., 2008b. Political strategy and climate policy. *In*: Hugh Compston and Ian Bailey, eds. *Turning down the heat: the politics of climate policy in affluent democracies*. Basingstoke: Palgrave Macmillan, 263–288.

Dolzak, N., 2001. Mitigating global climate change: why are some countries more committed than others? *Policy Studies Journal*, 29, 414–436.

Ereaut, G. and Segnit, N., 2006. *Warm words: how we are telling the climate story and how we can tell it better*. London: Institute for Public Policy Research.

Giddens, A., 2009. *The politics of climate change*. Cambridge: Polity.

Hajer, M. and Veersteeg, W., 2005. A decade of discourse analysis of environmental politics: achievements, challenges, perspectives. *Journal of Environmental Policy and Planning*, 7, 175–184.

Helm, D., ed. 2005. *Climate change policy*. Oxford: Oxford University Press.

IPCC (Intergovernmental Panel on Climate Change), 2007a. *Climate change 2007*. Cambridge: Cambridge University Press.

IPCC, 2007b. Summary for policymakers. *In: Climate change 2007: mitigation. Contribution of Working Group III to the Fourth Assessment Report*. Cambridge: Cambridge University Press.

Kerr, A., 2007. Serendipity is not a strategy: the impact of national climate programmes on greenhouse-gas emissions. *Area*, 39, 418–430.

Newell, P. and Paterson, M., 1998. A climate for business: global warming, the state and capital. *Review of International Political Economy*, 5, 679–703.

O'Riordan, T. and Jäger, J., 1996. *Politics of climate change: a European perspective*. London: Routledge.

Oshitani, S., 2006. *Global warming policy in Japan and Britain: interactions between institutions and issue characteristics*. Manchester: Manchester University Press.

Rabe, B., 2004. *Statehouse and greenhouse: the emerging politics of American climate change policy*. Washington, DC: Brookings.

Stern, N., 2007. *The economics of climate change, the Stern review*. Cambridge: Cambridge University Press.

Tans, P., 2009. *Recent monthly mean CO_2 at Mauna Loa (2005–2009)* [online]. Available from: http://www.esrl.noaa.gov/gmd/ccgg/trends/ [Accessed 27 April 2009].

Risk analysis and climate change

Nick Pidgeon and Catherine Butler

There is an increasing emphasis on risk-based approaches in the scientific and economic assessment of climate change, exemplified by the Stern Report and IPPC 4th Assessment. In the United Kingdom, risk discourse also increasingly dominates environmental policy-making and governance. The use of risk assessment, management and communication practices in climate change governance and policy is critically examined, utilising an interpretation of 'risk' as a knowledge practice for informing decision-making and an instrument for governing populations. In elucidating current risk practices, alongside key critiques and varied proposals for revised approaches to risk characterisation, both the capacities *and* limitations of a risk basis for policy aimed at delivering adaptation and deep cuts in greenhouse gas emissions are examined. While contemporary risk approaches align well with dominant political rationalities in affluent Western democracies, they have serious limitations as a basis for the delivery of aggressive climate policy aims.

Introduction

The problem of climate change sets the science and policy communities a range of complex challenges, many involving questions of uncertainty and risk. While human influence upon the world's climate is now widely accepted, detailed climate projections depend upon the characterisation of imperfectly under-stood and complex physical environmental systems in interaction with equally unpredictable social processes. To be timely and effective, policy-making for mitigation and adaptation will require decisions in relation to a future holding physical and social unknowns with little time to make detailed projections or wait until better information is available. In this context, recent years have seen an increasing emphasis on risk approaches in the scientific evidence for informing political decision-making, and in the wider governance strategies being adopted.

Fundamental climate science has always acknowledged complex uncertainties in its models (Oppenheimer 2005), although policy documents until quite recently have tended to avoid explicit risk-based predictions. The 2001 Intergovernmental Panel on Climate Change (IPCC) assessments utilised narrative-based scenarios for dealing with future uncertainties, but did not assign subjective probabilities to scenarios. By contrast, the fourth IPCC (2007) assessment report (AR4) adopts uncertainty expressions (medium confidence, highly likely, etc.) throughout its analysis, while the UK's 2006 Stern (2007) review of climate change economics advocates the assessment of risk and uncertainty for decision-making. As a governance tool, risk discourse has proliferated as a means of framing and managing social issues, and now extends into climate policy approaches to both adaptation (DEFRA 2005) and mitigation (DEFRA 2008).

We seek to understand the role that analytic tools and governance strategies derived from risk and uncertainty analysis might play in climate policy. We take a critical stance towards this question by outlining some of the key elements of a risk-based approach as well as varied critiques of its use. Our central argument is that while in principle formal risk methods constitute an important suite of tools for informing decisions about climate policy and evaluating options (see Lorenzoni *et al.* 2005), in practice climate change exhibits, *par excellence*, some of the very characteristics that render conventional risk analysis tools problematic. These include such things as scientific and social complexity, deep forms of uncertainty and ambiguity, debates about incommensurable values, temporal and spatial inequalities, systemic causes, and governance dilemmas. Consequently, it remains an open question whether risk analytic approaches to decision-making and governance can genuinely help policy-makers deliver appropriate adaptation measures and deep cuts in greenhouse gas emissions.

The politics of risk and the risks of climate change

Risk characterisation – basic elements

In modern dictionary terms, risk refers to 'danger, (exposure to) the possibility of loss, injury or other adverse circumstances' (Hacking 2003, p. 24). This basic definition does not, however, address the myriad ways in which risk is treated in conceptual thought and practice (Pidgeon *et al.* 1992). While the concept of risk dates back at least to the sixteenth century, formalised analytic practices are more recent. Hacking (2003) specifies the late 1960s as the era in which the *professional practice* of risk analysis became an institutionalised activity, while the birth of modern risk–benefit thinking is often attributed to Chauncey Starr's (1969) classic paper in the journal *Science*. Early applications included evaluations of industrial safety, the risks of spaceflight and of nuclear power.

Risk is typically viewed in the technical literature as some quantitative combination (usually the product) of the likelihood and consequences associated with a hazard, with generic approaches often characterised by four elements:

(a) *Problem Framing.* Structuring the problem: identifying key hazards and their characteristics, the extent of the system being modelled, consequences and likelihoods to consider, key uncertainties or gaps in knowledge, and the impact of social systems on modelling.

(b) *Risk Assessment.* Quantitative evaluation: of evidence and judgements regarding probabilities and consequences; mathematical integration of model elements; and judgements regarding the acceptability or tolerability of risks.

(c) *Risk Management.* The development of policy options and appraisal against broad objectives, balancing costs and benefits, decision implementation and monitoring.

(d) *Risk Communication.* Various forms of (one-way or two-way) communication with stakeholders and affected parties for numerous purposes, including structuring the problem, eliciting values and subjective likelihoods, communicating or deliberating results.

Simple enough to define in theory, the distinctions above are often difficult to maintain in practice. Separation of assessment and management is particularly problematic, while communication and two-way stakeholder engagement is now seen as integral to *all* stages of the assessment and risk governance process rather than something to be adopted separately after risk assessment and management issues have been resolved (Stern and Fineberg 1996).

Contemporary environmental management also encompasses risk–benefit and cost–benefit analysis. Jacobs (1991, p. 196) explains that 'cost–benefit analysis starts from a simple premise, namely that an investment project should only be undertaken if its benefits outweigh its costs'. Risk assessment activities thus intertwine with the calculation of (primarily economic) costs of probable outcomes, while the weighing of costs and benefits forms part of the basis for value judgements about risk and its management.

Beyond the formalised definitions and processes of mathematical risk analysis, social theorists have examined risk practices in terms of their wider implications. Within the governmentality tradition, risk is examined as a way of thinking, a rationality that is seen to take multiple forms. From this perspective, risk rationality is seen as integral to contemporary forms of governance (Dean 1999). New regimes for risk management transform aspects of social life such as social integration, value-based reasoning, morality, self-identity and governance. Risk analysis – in this broadest sense – should be treated as *both* a pervasive technology for governing populations and as an analytic tool for informing decision-making and policy.

Risk thinking in UK policy and politics

Since the 1990s risk analysis, management and communication have been normalised in UK institutional discourse (Power 2004). The publication of the Strategy Unit Report *Risk: Improving Government's Capability to Handle Risk and Uncertainty* (Cabinet Office 2002) provides one example of the embedding of risk talk in policy processes, with 'handling risk' seen as central to the work of government because of a perceived qualitative change in uncertainties brought about by the rapid pace of technological and scientific development. Multiple examples of the embedding of risk talk can be found across UK government departments, in associated policy documents, and in the establishment of policy teams to support the embedding of risk practices (see, for example, HM Treasury 2004, DEFRA 2008).

A feature of the contemporary expansion of risk talk in the UK is the *individualised* form that it now takes (Rose 1999). Drawing on the work of Foucault, risk management is seen not only as a means for evaluating problems and deciding among policy alternatives but also as a particular way of governing populations. Risk is seen as 'a particular style of thinking born in the nineteenth century' that entailed the disciplining of populations 'by means of the statistical intelligibility that the collective laws of large numbers seemed to provide' (Rose 1999, p. 247). Dean (1999, p. 191) notes that a 'fundamental tenet of the governmentality literature ... is that calculative rationalities such as those of risk have a certain political polyvalence i.e. they can be invested with different sets of purposes depending on the political programme and rationalities they come to be latched on to'. In advanced affluent democracies the *individualisation* of risk is a specific representation that can be seen as linked to new forms of liberal government (Dean 1999, p. 191). The collectivisation of risk in forms of social insurance, such as the welfare state, which spread risk across populations, is increasingly being displaced as management becomes individualised. Individuals, communities, families and organisations are being urged by politicians to *take upon themselves* responsibility for social issues. In contemporary UK policy, this form of individualised risk management becomes a policy solution to multiple issues. Given the proliferation of risk, both as a knowledge practice and as a particular form of governance, it is unsurprising that risk tools have begun to be adopted as political strategies, as opposed to merely technical ones, in relation to mitigating and adapting to climate change.

The emergence of risk in climate policy

A number of developments in climate policy point to a gradual incorporation of risk-based thinking. The publication of AR4 in 2007 marked a distinctive shift away from the scenario-based approach adopted previously in providing both uncertainty and confidence estimates to accompany its assertions (IPCC 2007). The report commences with a note on 'treatment of uncertainty' in which it distinguishes between three variants: (1) uncertainty expressed qualitatively through assessment of a balance of evidence (medium evidence, much evidence); (2) uncertainty assessed more quantitatively through subjective expert judgement of confidence in underlying models and evidence; and (3) quantitative likelihoods (explicitly defined as numeric probabilities) based upon a combination of judgement and statistical modelling. Using this typology the report frames its conclusions using a series of uncertainty statements, for example stating that the observed increase in global average temperatures since the mid-twentieth century is *very likely* (defined as probability >90%) due to the observed increase in anthropogenic greenhouse gas concentrations; that the negative impacts of climate change on freshwater systems outweighs its benefits (*with high confidence*); and that there is *high agreement* and *medium evidence* that changes in lifestyle and behaviour patterns can contribute to climate change mitigation across all sectors of the global economy.

The parallel Stern report (2006) on the economics of climate change also explores the implications of adopting such an approach for decision-making. Stern argues that multiple uncertainties inevitably pervade any economic modelling that we might undertake. Taking an approach to cost–benefit risk analysis that is markedly different from conventional economic thinking, a point to which we return later, Stern argues that any economic analysis 'must be *global*, deal with *long* time horizons, have the economics of *risk and uncertainty* at its core, and examine the possibility of major, *non-marginal* changes' (2007, p. 25, emphasis original). A fundamental conclusion of his analysis is that climate change poses a significant risk of potentially catastrophic impacts upon global society and that the less costly option is to intervene early.

Both AR4 and Stern have been important for informing international policy approaches to mitigation and adaptation. A further development is the United Kingdom's Climate Impacts Programme (UKCIP) which, in collaboration with the Met Office and Hadley Centre, aims to provide *risk-based* estimates of the regional impacts of climate change for adaptation purposes, while the independent Climate Change Committee set up by the UK government in 2008 is also committed to developing risk-based approaches. In addition, the UK's adaptation policy for the climate-related risk of flooding already entails a 'new' risk-based approach. In a House of Commons enquiry following major floods in 2007, the 'risk-based approach' was cited by the Environment Agency's Chief Executive, Baroness Barbara Young, as a major part of the changing way that flood management is delivered in the context of climate change (House of Commons EFRA 2008), the implication being that the development of predictive risk knowledge represented one of the major policy changes needed to deliver adaptive strategies. In this sense risk calculative rationality is invoked not only as a benign form of knowledge but also as an integral part of the *political* strategy for adapting to climate change (Butler 2008).

We also find the uptake of 'risk as governance' in relation to mitigation policies. In her examination of the cities for climate protection campaign, Slocum (2004) has asserted that the particular focus of emerging strategies (for example energy efficiency, education, local benefits) was bound up with neoliberal political rationalities. In particular, climate change, as constituted through the lens of cost–benefit analysis, becomes an economic problem, with particular consequences for the kinds of approaches and policies that are pursued. While this study was located in the US, similar findings are evident in UK contexts (see, for example, Bulkeley and Kern 2006).

A further means through which risk thinking and governance appears to be becoming entwined in climate change policy is the political focus on individualised risk management of climate risks. Individualised everyday actions, such as leaving electrical equipment on standby and turning your thermostat up, have been positioned in relation to their role in causing global climate change (DirectGov 2008), thereby situating citizens as individually responsible risk managers. This can be characterised as representing a form of governance whereby citizens are ruled through their freedoms and choices, with responsibilities to make the *right* choice designated through communicated policy messages and approaches. This aspect of climate policy, whereby emissions are constituted as the responsibility of individuals, can be construed as a political manoeuvre in the face of an inherently difficult and uncertain issue that among other things deflects blame away from political institutions when things do not turn out as planned (see Hood *et al.* 2001). As such, there is nothing unique about risk when applied in this way: risk technologies are merely particular *scientific tools* through which societal issues are constituted as governable. The focus on changing individual behaviours as a means of mitigating climate change can thus be seen as part of a wider set of shifts through which social issues are constituted as being manageable through individualised risk management aligned with advanced liberal political rationalities. In the UK, then, risk analytic processes and governance strategies underpinned by risk rationality are becoming increasingly intertwined with (and part of) political strategies for addressing climate change.

Interpreting the proliferation of risk: a policy tool which 'fits'

In understanding the pervasiveness of risk practices in climate change governance, we propose that it is necessary not only to delineate the qualities of risk methodologies but also to understand the role that societal, governmental and structural processes play in maintaining and furthering the application of 'risk' in policy. In this section we delineate three key 'aspects of alignment' through which risk 'fits' with wider contemporary societal processes: 1) risk rationality fits with current advanced liberal political forms of government; 2) risk practices provide important knowledge artefacts in an increasingly uncertain world; and 3) risk analytic practices meet the requirements of economic rationalities in capitalist societies.

A number of scholars have pointed to shifts in forms of governing that are characteristic of, and complementary to, the organisation of advanced liberal democracies (Rose 1999). From this perspective, risk is analysed as a component of assemblages of practices, techniques and rationalities concerned with how we govern. Risk is a calculative rationality that is tethered to assorted techniques for the 'regulation, management and shaping of human conduct in the service of specific ends' (Dean 1999, p. 178). Notions of risk are thus 'made intelligible as specific representations that render reality in such a form as to make it amenable to types of action and intervention' (Dean 1999, p. 178). The strategies of government associated with advanced liberal rule seek to utilise and instrumentalise the possibility for *action at a distance*: government without coercive or strong regulatory intervention. In contemporary governance the collectivisation of risk (within the social state) is displaced and instead individuals, families, organisations and communities are called upon to take responsibility for and manage an ever increasing number of risks. As such, risk technologies provide one means of interpretation that renders reality amenable to 'the desire to govern at a distance' (Rose 1999, p. 43). From this perspective, risk approaches to governance and decision-making may be dominant precisely because they can be aligned with advanced liberal political rationalities.

Second, the popularity of risk knowledge practices may be a reaction to the heightened uncertainties associated with contemporary governance. It is debateable whether the world is more risky now than in the past, but less controversial to suggest that 'more possible outcomes in the world are regarded as amenable to human intervention and decision, rather than being in the hands of the Gods' (Power 2004, p. 14). In this sense the pressure to make decisions (when outcomes are seen as the product of human decision-making) can be seen to drive a focus on risk rationality as the modern western world's dominant tool for gaining knowledge of the future and coping with uncertainties; a function which in the contemporary world has come to be seen as 'central to the business of government' (Cabinet Office 2002, p. 4). Risk discourse implies control, manageability and accountability. While, as we indicate below, such notions may be inappropriate for complex issues like climate change, they offer an illusion of governance in that they construct inherently difficult issues as both tractable and manageable.

Finally, in contemporary capitalist societies profit-making and economic growth continue to be significant aims of government, with economic cost–benefit analysis, underpinned by the doctrine of classical economics, offering important ideological and practical tools towards achieving this end. Oels (2005) asserts that the scope for policy interventions in relation to climate change is restricted by liberal political rationalities that favour market-based solutions, with legitimacy for action determined by cost–benefit analysis rather than, for example, any moral imperative. The complexities involved in shifting away from classical economics to different modes of economic-based decision-making, or of changing economic arrangements more broadly, are beyond the scope of this paper. However it is important to point to the significance of cost–benefit analysis (as part of the risk instrument) in political contexts where capitalist interests are of importance for decision-makers. This 'fit' with mainstream economic thinking adds a final reason for the proliferation of risk approaches across multiple areas of contemporary policy. We suggest that understanding these wider relations, and their significance in furthering the application of risk approaches to policy problems, is of importance for examining the adoption of risk in climate policy.

Despite considerable rhetoric, UK policy and political decision-making relating to climate change has yet to yield deep cuts in greenhouse gas emissions (Lorenzoni *et al.* 2008). Our argument is that risk rationalities alone are unlikely to enable the delivery of effective climate change adaptation and mitigation, while at the same time their degree of fit with existing socio-political conditions is playing an important role in their uptake. In this sense, the development of any strategy for delivering deep cuts in greenhouse gas emissions must pay attention to the *relations between* existing political, economic and social conditions and the proposed approach. In the following section we delineate key aspects of literatures that critically address the uses of risk analytic practices in relation to complex socio-environmental and technical issues, and thereby open up a different perspective on the uses of risk in climate policy contexts.

Critique of risk-based approaches

Critique of risk I: technical

Writing from a decision analysis perspective, Morgan *et al.* (1999) highlight the limitations of conventional risk policy tools when applied to global environmental change issues, criticising in particular key assumptions. First, they assert that there is an inbuilt assumption 'that there is a single public sector decision-maker who faces a single problem in the context of a single polity' (p. 271). They point out that this assumption fails to recognise the institutional and political complexities involved in most climate decision-making, where multiple parties are involved across a range of complex and interdependent political domains. The second critique concerns the assumption that impacts on society will be small compared to the resources available to deal with them, something which is integral to the economic valuation of impacts. In developed nations the costs of climate impacts may well be 'small', but for developing countries these will be proportionally far greater, making conventional calculation problematic. In addition, impacts for *all* nations may be far larger than previously thought (Stern 2007), while the possibility exists that major destabilisations (tipping points) in the climate system might lead to rapid and catastrophic global changes (Lenton *et al.* 2008).

Morgan *et al.*'s (1999) third critique concerns the assumption that values are known, and exogenously determined. For environmental 'goods' which are not traded in any market, such as biodiversity, economic analysis typically relies upon statements of individual preference to arrive at valuations. But psychological evidence now suggests that elicitation of preferences for environmental goods faces conceptual and methodological challenges (Fischhoff 1991, Satterfield 2001). Individual preferences are *constructed* (Lichtenstein and Slovic 2006) rather than being inherent to individuals, or in some other way exogenous and given. For Morgan *et al.* (1999) the issue lies not so much in the inherent difficulties in eliciting climate-related preferences as in the lack of past experience with climate-related values. They argue that such disadvantages can be overcome by the incorporation of uncertainty about future values into analyses, the selection of strategies that preserve future flexibility, and directly incorporating wider considerations (such as inequity and justice) in the definition of utility.

Morgan *et al.*'s fourth point concerns the adequacy of economic discounting for describing time preferences. They do not take issue with the fundamental *principle* of discounting, suggesting that 'barring catastrophes ... wealthy future generations in the developed world will be able to deal with the impacts of global change without too much difficulty' (Morgan *et al.* 1999, p. 277). But, as noted above, recent analyses suggest that the possibility of catastrophe is not negligible, even for the developed economies, while other analysts, such as Jacobs (1991), criticise discounting for being inappropriate in contexts, such as the environment, in which non-market costs are high. Issues relating to the impossibility of measuring future people's environmental valuations (Adam 1998), and non-substitutable loss of natural habitat (Neumayer 2007), are also not addressed by conventional discounting techniques. Moreover economic discounting is predicated on the *assumption* of economic growth; as long as this is achieved the future is secure, as future generations become wealthier and better equipped to deal with impacts. In this sense analytical tools based upon cost–benefit analysis may be prone to unintended consequences, because the hidden value judgements underpinning the methodology will privilege political strategies that maintain economic growth.

Morgan *et al.*'s fifth critique addresses the representation of risk and uncertainty. They suggest that, in the context of climate science, 'traditional tools for analysis and display of uncertainty are inadequate' (Morgan *et al.* 1999, p. 278). AR4 (IPCC 2007) represents an important step forward in meeting this critique, but even this analysis fails to fully incorporate the potential for major future surprises (Schneider and Kuntz-Duriseti 2002) and ultimately the *epistemic ignorance* (Stirling 2006) inherent in our representations of both the physical and social systems underlying climate change processes. Morgan *et al.*'s final critique extends this argument by pointing to the difficulties in applying linear models to the non-linear systems associated with global change: in particular, key feedbacks and irreversibilities might be masked by analysis. Recent climate science is indeed starting to address such processes (Lenton *et al.* 2008), while Morgan *et al.* (1999) suggest that surprises can be approached by modesty in generalisations and the constant search for gaps in current analyses.

Seen from within the field of decision and risk analysis, both formal analysis and risk governance are desirable things in climate policy because they provide a basis for structuring and deliberating elements of policy decisions in a context that presents both complexity and uncertainty. Deficiencies of risk rationality can, in theory, be dealt with through the amendment and development of conventional risk tools, for example through approaches designed to reduce uncertainty, target critical knowledge gaps, or use precaution (Kasperson 2008), rather than requiring a wholesale transformation of approach. However, their deficiencies have been only partially addressed in recent developments in climate change policy, and more fundamental critiques of risk analysis as a policy instrument remain.

Critique of risk II: conceptual

Critiques from 'without' the risk analysis field contend that risk approaches to decision-making and governance break down under the circumstances that arise with climate change. These critical analyses vary in their focus but share a concern that the paradigm upon which risk rationality is grounded might be fundamentally inappropriate for comprehending and addressing complex global hazards.

In conceptual terms, Beck (1992) has been particularly critical of the application of risk rationality to global environmental hazards. For Beck, applying risk rationality entails an implicit depiction of such hazards as amenable to conventional procedures of calculation, management and control within the capacities of contemporary industrial society's established institutions. He argues that such hazards are, in actuality, characterised by temporal and spatial complexities which render conventional actuarial calculation problematic and create severe limitations for control and management. For Beck (1992, p. 22), 'dealing with the consequences of modern productive and destructive forces in normal terms of risk is a false but nevertheless very effective way of legitimising them'. In this sense, risk rationality facilitates neglect of the *systemic* nature of contemporary environmental and technological hazards (Turner and Pidgeon 1997). Writing from a temporal perspective, Adam contends that socio-environmental risks such as climate change are typically discussed in terms of 'scientific proof, certainty, prediction of the future based on knowledge of the past, risk calculation, and safety in the normal sense of the word' (Adam 1998, p. 37). She designates this as a seemingly ill-fitting conceptual package for dealing with issues characterised by immanence, (manufactured) uncertainty, invisibility and extensive latency.

Wynne (2002) similarly takes a critical stance on risk in environmental policy, and on the dominant scientific-institutional risk culture in the interpretation and appraisal of socio-environmental and technological issues. For Wynne (2002, p. 460), risk claims are 'endemically and increasingly contested', reflecting not only 'mere uncertainty in propositional claims about consequences' but also a fundamental epistemological ambiguity associated with *realist environmental and risk discourse*. The failure to recognise this ambiguity is for Wynne problematic, with the continued efforts to lament, purify and delete the ambiguities and 'unacknowledged human discourses which shape, and are projected by, the institutional scientific policy discourses of risk' being detrimental both to democratic and effective decision-making (Wynne 2002, p. 460).

Within the governmentality tradition, risk discourse has been positioned as one component of 'assemblages of practices, techniques and rationalities concerned with how we govern' (Dean 1999, p. 132). The critical perspectives that this line of thought offers are diverse and applied to differing forms of risk rationality. In a general sense, this tradition positions risk knowledge as one specialised language and set of practices through which power is channelled. Risk rationalities implicitly empower some as experts and exclude others, while also acting to shape and govern citizens in particular ways. Individualised risk management has been criticised as a corollary of the inability of institutions to tackle complex contemporary hazards through existing legal, political, economic and scientific systems; and as merely a means for diverting accountability while facilitating the *continuation* of existing arrangements and of the hazards produced through them. Engaging with complex issues like climate change in terms of risk, then, masks the true limitations of current institutions and analytic approaches for interpreting and managing global hazards.

The above critiques contribute to our understanding of the limitations of risk approaches for tackling climate change but more crucially serve to challenge its effectiveness in political governance terms. One important implication is that the limited successes in producing deep cuts in greenhouse gas emissions and adapting to climate change evident thus far in the UK are unlikely to be improved through risk approaches in decision-making and governance. What these critiques do suggest is the need for different kinds of knowledge practices in political decision-making for climate change, practices designed to overcome some of the problems of risk rationality.

New approaches to uncertainty and governance

Beyond efforts to adapt risk analysis tools, various alternative knowledge practices to replace or supplement risk rationality in contexts of high uncertainty and complexity have been proposed. Such proposals often entail moves to facilitate inclusion of different knowledges in governance processes.

Reviewing the field in the mid-1990s, the US National Research Council (NRC) (Stern and Fineberg 1996, Dietz and Stern 2008) produced a set of proposals for risk characterisation which they term an *analytic-deliberative* process. Both welcomed and criticised as representing a clear conceptual break from former analytic practices and the over-dominance of experts in risk assessment and management, these proposals seek to combine risk modelling, uncertainty analysis and expert judgement with extended stakeholder deliberation *at all stages* of the risk characterisation process. Likewise, the International Risk Governance Council (IRGC) advocates a new approach to risk governance that offers different levels of analysis and stakeholder involvement (Renn and Walker 2008). It remains to be seen whether these proposals transcend existing limitations of risk tools in ways which are both intellectually coherent and workable in practice.

Funtowicz and Ravetz (1993) posit a new form of knowledge practice for responding to complex risk issues. They wish to maintain that scientific knowledge is important, arguing that conventional tools are most appropriate in situations characterised by low systems uncertainty and restricted decision stakes. But in contexts of high systems uncertainty, and where decision stakes are high, they call for a shift to 'post-normal' science to address the limitations associated with conventional analytic tools. They propose an extension of the peer community in the assessment of scientific knowledge and the inclusion of other knowledge(s) in decision-making processes.

Jasanoff and Wynne (1998) have been critical of Funtowicz and Ravetz (1993) for apparently assuming that extension of the peer community (including lay audiences) will lead to greater certainty for decision-making. They suggest that 'the implicitly forward looking storyline of the ... model echoes the idea of knowledge acquisition as perpetual progress' and contrasts with accounts that characterise the linkages between scientific knowledge production and politics as being complex and unpredictable (Jasanoff and Wynne 1998, p. 13). They, however, make related assertions regarding the need for involvement of a wider range of actors in public science issues, but from a different perspective. For Jasanoff (2002), risk assessment methods incorporate assumptions about physical and biological systems, as well as about human lifestyle and behaviour. Further, they contain built-in presuppositions and value choices that are hard to uncover, resulting in the setting aside of unresolveable cultural differences in people's conceptions of risk. As such, risk analyses fall short in the attempt to cope with incommensurables, such as risk to a sacred habitat or communal way of life, and the kind of indeterminacy that results from human interactions with nature. Accordingly, Jasanoff argues that, in the face of complex multi-vocal risk debates, decision processes which can incorporate the free interaction of different frames and new values are the means through which these kinds of issues can be better addressed.

Similarly, Wynne (1996, p. 39) proposes institutional reform involving 'recognition of new socially extended peer groups legitimated to offer criticism ... from beyond the confines of the immediate exclusive specialist science peer group'. Wynne (2002, p. 463), however, is critical of emergent attempts in policy cultures to develop scientific knowledge practices in more participatory directions (such as the NRC and IRGC proposals), as he believes that these have 'perversely reinforced attention only on the back-end scientific questions about consequences of risks (reflecting an embedded implication of prediction and control)'. For Wynne, contemporary efforts to address the issues he identifies in the institutionalisation of risk culture have, thus far, been inadequate.

All of the above envisage a role for lay public(s) in decision-making about complex issues. Beck (1992) retains more focus on *expert* knowledge, but calls for a reformation of this knowledge through the development of a reflexive scientisation. This involves the application of 'organised scepticism' *to science itself* as a possible means of identifying otherwise unanticipated consequences of its application. The focus is thus on *expert critique of expertise*, although a role is envisaged for citizen's groups in research policy decision-making processes.

Generally, these new approaches focus on the centrality of risk rationality in current political processes; the limitations of this, as a set of rationalities, and of wider *conventional* scientific authority; the politicisation of knowledge (that is, a blurring of the distinctions between values and facts); and the resultant requirement for a democratisation of approach so that different rationalities might be included in decision-making.

Changing climate, changing knowledge, changing practice

Despite increasing uptake of risk thinking in climate change governance and decision-making, societal changes which provide for deep cuts or successful adaptation have yet to be delivered. The critiques discussed above provide insight into some of the reasons for this failing. Efforts to address critiques of risk-based thinking are evident in proposed alterations to risk tools and more fundamental changes in knowledge practices or institutional style. Such shifts in knowledge practices for climate policy decision-making processes are evident in the UK. These include the 2007 citizen's climate change summit convened to incorporate the views of citizens into climate policy, as well as moves to develop 'expert' and 'value-based' uncertainties in integrated assessment models (Dessai *et al.* 2004). But these have yet to result in different decision outcomes or stronger political action.

We suggest that attempts to shift to new knowledge practices, or to adapt conventional risk analytic tools for climate change policy, are limited – in the UK at least – by the wider socio-political contexts in which knowledge is applied. This is evident in examples pertinent to climate change policy. First, in relation to adaptation policies for flooding, and despite efforts to include wider aspects of value (including social and environmental costs), *economic* assessment takes priority. For example, the economic costs of any flood management scheme must represent value for money in the eyes of Treasury (DEFRA 2006). Efforts to adapt cost–benefit analytic tools to incorporate qualitative valuation do not entail a fundamental challenge to existing socio-economic and political contexts and systems, with the result that economic utility is privileged over other important (but non-economic or intangible) concerns. The primacy of economic assessment is tied to advanced liberal political rationalities and contemporary economic culture. Attempts to prioritise economic concerns over other aspects of value, and to convert all utilities to economic value, are thus *socially and institutionally ingrained* and not purely methodological choices.

A second example is the application of individualised risk management techniques in climate change governance, which sees individual citizens targeted for their roles in causing greenhouse gas emissions and urged to take responsibility for their own emissions. The locus of responsibility is thus shifted from government to individuals despite emerging evidence of public concern and calls for the obverse of this: strong political action and leadership on climate change. This has considerable potential to lead to a 'governance trap' in climate policy, as ordinary people call for strong action by governments (reasoning, quite rightly, that climate change is too large a problem for them alone to tackle) while governments, afraid to commit to tough decisions in the face of the electoral cycle, seek to place responsibility for risk management upon the shoulders of individuals (Lorenzoni *et al.* 2008). This form of governance is closely aligned with the political rationalities associated with both ecological modernisation and advanced liberal democracies, namely rule at a distance and through markets. In such a context, alignment with the formula of political rule appears likely to take priority over the practical effectiveness of the strategy employed. Additionally, at precisely the moment when anthropogenic causes are beyond doubt, undermining the bases for any lingering climate scepticism, the imposition of ostensibly more 'rational' risk-based approaches to characterise impact uncertainties (as in the new UKCIP scenarios) might have the unintended effect of *reducing* people's willingness to change their behaviour, since if some elements of the problem remain uncertain, why bother to alter our lifestyles?

Third, the Stern report (2006) entailed efforts to adapt conventional cost–benefit and risk analyses in ways more appropriate for application to climate change. As Weitzman (2007) succinctly points out, the conclusions of the Stern report were at odds with most conventional economic analyses of the climate change issue. Stern highlights the fundamental ethical issue underlying climate decision-making at a global level: that of intra- versus inter-generational equity. In order to deal with this issue (and as an important part of its novelty) the Stern analysis selects a very low discount rate for future damages. However, he also incorporates the possibility, suggested by some of the most up-to-date climate modelling, that climate sensitivity (the expected mean temperature rise under a doubling of pre-industrial greenhouse gases) could be much higher than originally thought. In particular, while typical estimates of sensitivity lie somewhere in the range 2–3°C – bad enough in terms of impacts, particularly for already vulnerable species and habitats, as well as people in the developing world – there is a non-zero but uncertain probability that it will be greater (of the order of 5–10°C). The damages under such an uncertain but 'fat tail' scenario might be so catastrophic for the world as to dominate the entire climate decision problem. As well as challenging the received economic analysis and thinking on the matter, the potential for catastrophic outcomes, however unlikely, might outweigh the effects of even conventional economic discounting.

The Stern assessment has had a significant impact on international political discourse and thought regarding the need for mitigation. It has proved a powerful ally to the environmental movement in asserting coherently and comprehensively *in economic language* the need for aggressive emissions reductions. The successes of Stern can be seen as linked to its alignment with wider economic priorities, as it positions the climate change issue in such terms. The effects of the Stern report should not be overstated, however, as UK emissions cuts remain limited and political strategies for reduction continue to be conservative, while Stern's adoption of the lowest possible discount rate, in order to address the problematic assumptions that conventional discounting brings, has been subject to methodological critique precisely because of its departure from conventional analysis (Neumayer 2007, Nordhaus 2007). In a related line of enquiry, Slocum (2004) has been critical of focusing climate mitigation policy through an economic lens as a pragmatic means of getting climate change onto political agendas. In her research examining the cities for climate protection campaign she asserts that while the particular emphases of the campaigns may be seen as a pragmatic response to contemporary political contexts, they also amount to 'a neoliberal buffet of options that do not address values nor necessarily lend themselves to structural change' (Slocum 2004, p. 772). This suggests the need to be wary of approaches that fit with existing rationalities if they only provide limited scope for change.

Finally, Lorenzoni *et al.*'s (2007, p. 67) analysis of UKCIP positions it as an example of a 'boundary organisation' created specifically 'to enhance society's capacity to respond to climate change' as a *post-normal* issue. They position 'boundary organisations' as facilitating 'cooperation to achieve a shared objective or the "co-production of mutual interests" by a) creating scientific and social order ... [and] b) acting as agents of both politicians and researchers'. They explain that UKCIP 'lies at the interface of various perspectives on climate change, fostering practical responses to short-term and medium term climate impacts' (Lorenzoni *et al.* 2007, p. 70). As such, the operational characteristics of UKCIP 'appear to be hallmarks of a post normal science approach', indicating that it operates 'as an intermediary between science, policy and politics' (p. 70). They note, however, that their research revealed difficulties associated with constraints placed on such new institutional forms by the necessity for them to 'function within, and [be] subject to, the current system' (p. 68). Constraints were identified relating to stakeholders' perceptions of climate change and uncertainty; the paucity of legislative and regulatory guidance; lack of funding; absence of useful best practice examples; and limited business interest in the main goal of adaptation. Further tensions were explicated around the dilemma of maintaining independent research whilst being policy relevant. In many of these examples the constraints are externally defined and indicative of wider limitations on the capacities of existing institutions to adopt post-normal approaches (Lorenzoni *et al.* 2007). Difficulties in adopting more radical approaches to climate change, as in the development of post-normal institutional forms, may arise precisely because they no longer fit with wider socio-political and economic rationalities.

These examples provide insight into both the limits of conventional risk approaches for addressing climate change and the difficulties that arise in adapting these and in applying new approaches in efforts to advance climate policy. They reveal the interplay between risk, as a particular knowledge and governance practice, and the wider conditions with which it aligns, as well as providing insights into the difficulties which may arise in instituting new practices without also addressing the wider socio-political and economic context.

Concluding comments

There has been a gradual incorporation of risk-based approaches into climate science and policy. This evolution can be seen as linked to an embedding of risk rationalities in policy domains more generally. Of course, climate science and policy is inherently uncertain, and hence properly founded analytic approaches to uncertainty characterisation will be required if we are to move towards a post-carbon politics and economy. However, while existing risk approaches form an important suite of tools for analysing climate policy, in practice embedded assumptions, as well as built-in value choices and presumptions, render risk analysis problematic for addressing climate change. This alone should make us wary of uncritical acceptance of these methodologies and the wider rationalities which underpin them in the climate policy domain, and it remains an open question as to whether risk thinking as a basis for decision-making and governance can indeed help policy-makers to deliver deep cuts in greenhouse gas emissions and effective adaptation measures.

We have suggested that the alignment of risk approaches with existing socio-political, legal, scientific and economic rationalities in contemporary affluent democracies makes them potentially powerful agents for change, as exemplified in the significant influence of the Stern Report's call for aggressive mitigation. However, we also caution that, in relation to complex issues such as climate change, risk rationality may be inherently problematic. We have outlined some of the proposed roots for epistemic change. These entail either the adaptation of calculative risk tools (in the multiple forms that they take), to make them better suited to the task of interpreting climate change, or more fundamental changes to governance approaches and the incorporation of differing perspectives. Such shifts do, however, remain constrained by the wider structural and political contexts in which they are attempted. For this reason we contend that in order for novel strategies to be effective, the relations between them and the constraints associated with the socio-political-economic system will need to be addressed. That is to say, strategies may need to address changes to political rationalities and the social system in order to be effective.

In conclusion, risk analysis and its management holds an important place because of its current dominance and increasing application in climate change policy. However, as a set of tools it is unlikely to provide a basis for the delivery of effective adaptation or deep cuts in greenhouse gas emissions. Analysing why risk is an increasingly dominant approach in climate change policy operates as a forensic case study for understanding the potential for new strategies. In this sense, our conclusions hint at the need to develop policy strategies that address the structural and political contexts in which they are required to operate to ensure their effectiveness: strategies that facilitate changes in existing political rationalities. In practice, however, such strategies are likely to encounter multiple barriers to implementation, and as a result different approaches will not necessarily result in different decisions, and different decisions will not necessarily result in improved practices. Addressing the *relations and tensions* between all aspects of climate governance, and in particular presenting a challenge to many of the existing socio-economic and political assumptions underlying current environmental policies, will be crucial if we are to find and successfully apply ways of achieving adaptation and deep cuts in greenhouse gas emissions.

Acknowledgements

Preparation of this chapter was supported by the Leverhulme Trust (award F/00 407/ AG). We wish to thank Tim O'Riordan, Henry Rothstein, Hugh Compston and an anonymous reviewer for helpful comments.

References

Adam, B., 1998. *Timescapes of modernity: the environment and invisible hazards.* London: Routledge.

Beck, U., 1992. *Risk society: towards a new modernity.* London: Sage.

Bulkeley, H. and Kern, K., 2006. Local government and the governing of climate change in Germany and the UK. *Urban Studies*, 43 (12), 2237–2259.

Butler, C., 2008. Risk and the future: floods in a changing climate. *Twenty-first Century Society*, 3 (2), 159–171.

Cabinet Office, 2002. *Risk: improving government's capability to handle risk and uncertainty.* London: HMSO.

Dean, M., 1999. *Governmentality: power and rule in modern society.* London: Sage.

DEFRA (Department for Environment, Food and Rural Affairs), 2005. *Making space for water: taking forward a new government strategy for flood and coastal Management in England. First government response to the autumn 2004 making space for water consultation exercise.* London: Department for Environment Food and Rural Affairs Publications.

DEFRA, 2006. *Capital grants allocations for flood and coastal erosion risk management* [online]. DEFRA. Available from: http://www.defra.gov.uk/environ/fcd/policy/ grantaid.htm [Accessed 12 September 2008].

DEFRA, 2008. *Guidelines for environmental risk assessment and management* [online]. DEFRA. Available from: http://www.defra.gov.uk/environment/risk/eramguide/ [Accessed 7 March 2009].

Dessai, S., *et al.*, 2004. Defining and experiencing dangerous climate change. *Climatic Change*, 64, 11–25.

Dietz, T. and Stern, P., 2008. *Public participation in environmental assessment and decision-making.* Washington, DC: National Academy Press.

Direct Gov, 2008. *Directgov: ACT ON CO$_2$* [online]. DirectGov. Available from: http://campaigns.direct.gov.uk/actonco2/home/in-the-home.html [Accessed 7 March 2009].

Fischhoff, B., 1991. Value elicitation: is there anything in there? *American Psychologist*, 46, 835–847.

Funtowicz, S.O. and Ravetz, J.R., 1993. Science for the post-normal age. *Futures*, 25 (7), 739–755.

Hacking, I., 2003. Risk and dirt. *In*: R.V. Ericson and A. Doyle, eds. *Risk and morality*. London: University of Toronto Press.

HM Treasury, 2004. *The Orange Book: management of risk – principles and concepts.* London: TSO.

Hood, C., Rothstein, H., and Baldwin, R., 2001. *The government of risk: understanding risk regulation regimes.* Oxford: Oxford University Press.

House of Commons Environment, Food and Rural Affairs Select Committee (EFRA), 2008. *Flooding, Fifth Report of Session 2007–8, Vol. II (Ev10).* London: HMSO.

IPCC (Intergovernmental Panel on Climate Change), 2007. *Summary for policymakers in climate change 2007: The physical science basis. Contribution of Working Group I to the Fourth Assessment Report of the Intergovernmental Panel on Climate Change.* Cambridge: Cambridge University Press.

Jacobs, M., 1991. *The Green economy: environment, sustainable development and the politics of the future.* London: Pluto Press.

Jasanoff, S., 2002. Citizens at risk: cultures of modernity in the US and EU. *Science as Culture*, 11 (3), 363–380.

Jasanoff, S. and Wynne, B., 1998. Science and decision-making. *In*: S. Rayner and E.L. Malone, eds. *Human choices and climate change: the societal framework, vol. 1.* Columbus, OH: Battelle Press.

Kasperson, R.E., 2008. Coping with deep uncertainty: challenges for environmental assessment and decision-making. *In*: G. Bammer and M. Smithson, eds. *Uncertainty and risk: multidisciplinary perspectives.* London: Earthscan, 337–347.

Lenton, T.M., *et al.*, 2008. Tipping elements in the Earth's climate system. *Proceedings of the National Academy of Sciences of the USA*, 105 (6), 1786–1793.

Lichtenstein, S. and Slovic, P., 2006. *The construction of preference.* Cambridge: Cambridge University Press.

Lorenzoni, I., Pidgeon, N.F., and O'Connor, R.E., 2005. Dangerous climate change: the role for risk research. *Risk Analysis*, 25, 1387–1398.

Lorenzoni, I., Jones, M., and Turnpenny, J.R., 2007. Climate change, human genetics, and post-normality in the UK. *Futures*, 39, 65–82.

Lorenzoni, I., O'Riordan, T., and Pidgeon, N.F., 2008. Hot air and cold feet: the UK response to climate change. *In*: H. Compston and I. Bailey, eds. *Turning down the heat: the politics of climate policy in affluent democracies.* Basingstoke: Palgrave.

Morgan, M.G., *et al.*, 1999. Why conventional tools for policy analysis are often inadequate for problems of global change, an editorial essay. *Climatic Change*, 41, 271–281.

Neumayer, E., 2007. A missed opportunity: the Stern Review on climate change fails to tackle the issue of non-substitutable loss of natural habitat. *Global Environmental Change*, 17, 297–301.

Nordhaus, W., 2007. *The Stern Review on the Economics of Climate Change* [online]. New Haven, CT, Yale. Available from: http://nordhaus.econ.yale.edu/stern_050307.pdf [Accessed 12 August 2009].

Oels, A., 2005. Rendering climate change governable: from biopower to advanced liberal government? *Journal of Environment, Policy and Planning*, 7 (3), 185–207.

Oppenheimer, M., 2005. Defining dangerous anthropogenic interference: the role of science, the limits of science. *Risk Analysis*, 25, 1399–1407.

Pidgeon, N.F., *et al.*, 1992. Risk perception. *Risk-analysis, perception and management: report of a Royal Society study group*. London: The Royal Society, 89–134.

Power, M., 2004. *The risk management of everything: rethinking the politics of uncertainty*. London: Demos.

Renn, O. and Walker, K., 2008. *Global risk governance*. Dordrecht: Springer Verlag.

Rose, N., 1999. *Powers of freedom: reframing political thought*. Cambridge: Cambridge University Press.

Satterfield, T., 2001. In search of value literacy: suggestions for the elicitation of environmental values. *Environmental Values*, 10, 331–359.

Schneider, S.H. and Kuntz-Duriseti, K., 2002. Uncertainty and climate change policy. *In*: S.H. Schneider, A. Rosencranz, and J.O. Niles, eds. *Climate change policy: a survey*. Washington, DC: Island Press, 53–87.

Slocum, R., 2004. Consumer citizens and Cities for Climate Protection campaign. *Environment and Planning A*, 36, 763–782.

Starr, C., 1969. Social benefit versus technological risk. *Science*, 165, 177–196.

Stern, N., 2006. *Stern Review: the economics of climate change*. Cambridge: Cambridge University Press.

Stern, P. and Fineberg, H., 1996. *Understanding risk: improving decisions in a democratic society*. Washington, DC: National Academy Press.

Stirling, A., 2006. Precaution, foresight and sustainability: reflection and reflexivity in the governance of science and technology. *In*: J. Vos, D. Bauknecht, and R. Kemp, eds. *Reflexive governance for sustainable development*. Cheltenham: Edward Elgar, Publishing.

Turner, B. and Pidgeon, N.F., 1997. *Man-made disasters*. 2nd ed. Oxford: Butterworth-Heinemann.

Weitzman, M.L., 2007. The Stern Review on the economics of climate change. *Journal of Economic Literature*, XLV (September), 703–724.

Wynne, B., 1996. Misunderstood misunderstandings: social identities and the public uptake of science. *In*: A. Irwin and B. Wynne, eds. *Misunderstanding science? The public reconstruction of science and technology*. Cambridge: Cambridge University Press, 19–46.

Wynne, B., 2002. Risk and environment as legitimatory discourses of technology: reflexivity inside out? *Current Sociology*, 50, 459–477.

A strategy for better climate change regulation: towards a public interest orientated regulatory regime

Ian Bartle

The theory and analysis of regulation provides an understanding of climate change regulation and its limitations leading to the identification of political strategies to improve such regulation by orientating it more towards public interest objectives. Public interest and private interest theories of regulation are considered together with regime theory, which understands regulatory regimes as dynamic interactions of a variety of forces that can be altered by external pressure. One possible pressure is governmental action to increase regime transparency. UK and EU climate change regulation is used to consider whether increased transparency, in particular focused around a carbon price-based instrument with quantified emissions reductions, can shift the climate regime towards meeting the public interest. This runs up against difficulties of sociocultural plurality and the one-dimensional nature of the price mechanism. More successful strategy requires a price-based instrument coupled to transparent complementary policies and regulations that address sociocultural pluralities.

Introduction

Regulation of one form or another is the main mechanism for reducing greenhouse gas emissions and mitigating climate change. Climate change regulations vary widely. They include mandatory obligations or 'command and control' policies, financial incentives designed to induce behavioural change, and moral suasion to encourage behavioural and attitudinal change. The central regulatory mechanism in the EU is the emissions cap and trade system involving a combination of obligation (cap) and financial incentive (trade). Others include regulations which obligate emission limits for certain greenhouse gases, such as from landfill sites or industrial installations; financial incentives in the form of taxes or subsidies to favour renewable energy and conservation; and education and information campaigns to promote lower carbon lifestyles.

Climate change policies and regulations, however, have had limited success. The EU's cap and trade system initiated in 2005 has had a chequered start, notably owing to the over-allocation of national permits by governments under pressure from powerful industry interests (Baldwin 2008), and although the UK should meet its Kyoto obligations, the more onerous domestic emissions targets set in 2000 for 2010 will almost certainly be missed by a substantial margin.

My aim here is to show how the theory and analysis of regulation can provide an understanding of climate change regulation and its limitations that can lead to the identification of political strategies to improve such regulation.

The study of regulation has been complicated by the existence of different notions of the term. Three meanings can be distinguished which reflect varying concerns across disciplines within and beyond the social sciences (Jordana and Levi-Faur 2004, pp. 2–5): the first is a narrow sense of regulation as a set of authoritative or legal rules; the second refers to all actions by the state to control the economy and society; in the third and broadest sense, regulation consists of all forms of social and economic control whether by the state or by others.

Since the 1960s the study of regulation has been dominated by analyses and critiques of two major theories: public and private interest theories. These are sometimes referred to as 'economic' theories of regulation (Ogus 2004) and have focused primarily on the first meaning of regulation; that is, regulation as a set of legal rules.

Public interest theory is the usual entry point for the explanation of regulation because it explicitly addresses its purpose, which is to achieve publicly desirable outcomes when non-interventionary processes, notably the market, fail (Baldwin and Cave 1999). Although there are myriad 'market failures' across modern industrial society, climate change is so severe that it has been described as the 'greatest market failure the world has ever seen' (Stern 2007, p. xviii).

Public interest theory, however, has been criticised for decades, particularly by proponents of variants of private interest or 'capture' theory who stress that regulation derives from pressure of self-interested business groups who 'capture' regulators and legislators. This may explain why the record of climate change regulation in many countries is mixed at best.

Private interest theory also has empirical limitations, however, and the study of regulation in the last decade has moved away from these bipolar debates. Regulatory regime theory provides a more synthetic view and an empirically grounded way of explaining cross-regime variety and how regimes can change (Hood *et al.* 2001). Regime theory has not superseded private and public interest theories but involves attempts to understand regulation as a dynamic and multi-causal process.

Another significant turn in the study of regulation, sometimes labelled 'decentred regulation', parallels the 'governance' literature in which power is seen as less hierarchical and more distributed than in conventional models of government (Black 2001, Scott 2004). Regulation here is seen as more than a set of state-defined rules; it involves an understanding of regulation in its broadest sense, namely as a form of social control. There is no longer a sharp distinction between regulators (the state) and the regulated (the economy and society) (Scott 2004).

Furthermore, in the past two decades a range of techniques for devising 'good' or 'better' regulation have been developed by governmental practitioners. Perhaps the most prominent have been five general principles of good regulation articulated by the UK government: proportionality, accountability, consistency, transparency and targeting (BRTF 2003). It has been argued that these principles are not well reflected in emissions trading schemes (Baldwin 2008). However, one principle in particular, transparency, appears to offer significant potential to enhance governance and serve the public interest (Florini 2007) and thus has the potential to contribute to a strategy to improve climate change regulation.

This examination of regulation and the achievement of public interest outcomes in climate policy first outlines, in the next section, public and private interest theories and regime theory. The third section considers climate change regulation in relation to meeting public interest objectives, and particularly its failure to meet these objectives. The fourth section considers whether increases in transparency can shift the climate change regime towards the public interest, in particular by the adoption of a price-based instrument. The fifth section argues that a key weakness of the price instrument is that it involves a one-dimensional view of how individuals, organisations and governments think and act. It is therefore argued that regulations aimed at internalising market externalities using financial incentives need to be accompanied by policies which focus on other forms of behavioural change.

Public and private interests in regulatory theory

Public interest theory

Public interest theory of regulation is founded on the idea that the public interest is the justification for regulation and that this perspective provides the best way of understanding the form and existence of regulation (Ogus 2004). It suggests a certain altruism on the part of politicians and bureaucrats, in particular the 'trustworthiness and disinterestedness of expert regulators in whose public-spiritedness and efficiency the public can have confidence' (Baldwin and Cave 1999, pp. 19–20). The notion of the public interest is often broadly thought of as 'general well-being', reflects common values, and can be contrasted with sectional or vested interests. This broad understanding aligns with a definition of people as 'citizens', as opposed to consumers, who have a wide-ranging set of interests and values. This notion of the public interest can be contrasted with a narrower market-based idea in which people are defined primarily in relation to their economic interests (Feintuck 2004).

In the practice of regulation and the 'economic theory' of regulation, the market-based idea of the public interest has become dominant (Feintuck 2004, Ogus 2004). A 'market failure' approach to regulation, which refers to the constraint of 'public bads', has been seen as one of three main forces in regulatory regimes (Hood *et al.* 2001, p. 70). In the market-based approach the ideas of 'market failure' and 'market externalities' are inherent features (Baldwin and Cave 1999). Markets produce negative outcomes because certain factors are not considered in the market transaction itself. The production and consumption of a good can have a negative impact, for example on the environment, which is not reflected in the transaction. Regulation of one form or another can be introduced to 'internalise the externality'; that is, to ensure that unacceptable damage is avoided. A key goal of the public interest orientated economic theory of regulation is to strive for lowest cost forms of regulation (Ogus 2004).

The market-based approach has the advantage that the public interest is narrowly defined in the more tangible terms of the consumer interest rather than the vaguer and broader idea of citizens' interest. This is the more fashionable view of the public interest, but has the drawback of defining values in terms of economic interests and marginalising other values (Feintuck 2004).

Critique of public interest theory: private interest and capture theory

Despite the intentions of policy makers, regulation often fails to live up to public interest ambitions. Private interest, or 'capture' theory, offers a critique of public interest theory and purports to provide a more solid positive foundation for the understanding of regulation. It often commences with a critique of public interest theory as 'normative analysis as positive theory' (Viscusi *et al.* 1995, p. 326). Public interest and market failure might provide a normative rationale for what regulation should be, but it is less clear that it can explain what regulation is: it 'does not address the issue of how the potential for net welfare gains induces legislators to pass regulatory legislation and regulators to pursue proper actions' (Viscusi *et al.* 1995, p. 326). It has also been argued that the empirical basis of public interest theory is limited. Much research, particularly in the USA, has shown that many industries were regulated when there were no significant market externalities or natural monopolies (Viscusi *et al.* 1995, p. 326; and see Stigler 1971).

Key private interest theories stress that regulation derives from pressure exerted by self-interested business groups who 'capture' regulators and legislators. 'Capture theory', and its allies such as 'economic' and 'public choice' theories, dominated much of the literature in the 1960s and 1970s (Levine and Forrence 1990, Viscusi *et al.* 1995, pp. 327–339). They are based on the premises that the significant resource of the state is its power of coercion and that societal actors are rational and pursue actions that are utility maximising. Those societal actors who are best resourced and organised are the most able to convince the state to use its powers in their interests. Larger producers are generally the best placed, and private interest theorists argue that the empirical evidence shows that regulation is generally pro-producer. Private interest theories were dominant for many years such that one analyst noted that public interest approaches were 'left for dead' though they were 'found alive, although in a much weakened condition, in the 1980s' (Levine and Forrence 1990, p. 167).

Private interest theories have, however, also been subject to criticism on both empirical and theoretical grounds. There is, for example, much evidence of regulations that support less well organised and smaller interests, and a raft of social, environmental, health and safety regulations which were not supported by powerful private interests (Viscusi *et al.* 1995, pp. 328, 341). Trends towards liberalisation and deregulation in recent decades are not easily explained by the power of well-organised producers. In addition, the model of the political, policy-making and regulatory process in private interest theories may not be robust. It is not clear, for example, that interest groups have a substantial impact on elections or that legislators, once in power, are sufficiently threatened by loss of interest group support to implement policies favourable to these groups (Viscusi *et al.* 1995, p. 339).

The regime approach to regulatory formation

Public and private interest theories on their own therefore do not appear to provide a full explanation of regulatory outcomes. The regime approach to regulation offers the prospect of a more plural, synthetic and empirically grounded conceptualisation of regulation. It is based on the idea that regulation involves the interaction of multiple forces depending on circumstances and context. The notion of a 'regime' is established in political science and public policy analysis and usually denotes a plurality of processes embodied within a set of rules, norms and decision-making procedures. The term has been applied, for example, to 'risk regulation regimes' defined as a 'complex of institutional geography, rules, practice, and animating ideas that are associated with the regulation of a particular type of hazard' (Hood *et al.* 2001, p. 9). While less parsimonious than some theories, it offers the prospect of viewing regulatory regimes as 'systems'; that is, as 'sets of interacting or at least related parts rather than as "single cell" phenomena' (Hood *et al.* 2001, p. 9).

The interacting elements of the system can consist of forces and ideas, such as internalising market externalities or responding to interest group pressures. In a study of risk regulation regimes, Hood *et al.* (2001, pp. 61–69) focus on three elements. Firstly, responses to functional or market failures, which produce regulation that might be deemed as 'rational' and meeting the public interest. Secondly, 'interest-driven pressures', in which regulations reflect the interests of the best organised and politically most effective interest groups. Thirdly, 'opinion responsive pressures', in which regime content is shaped by broad public attitudes and preferences and varies depending on a wide range of social and psychological factors.

Empirical analysis of regulatory regimes reveals distinct cross-regime variation and change under external pressure (Hood *et al.* 2001, pp. 147–170). Understanding the dynamics of the regime can therefore enable policy makers to exert pressure for change. In any regime the question is how the balance of the regime can be shifted towards meeting public interest objectives. Prior to considering this, the next section outlines UK climate change regulation in relation to meeting the main public interest objective.

Climate change regulation and the public interest objective

A criticism of the public interest approach is that it is vague, particularly in how the public interest is identified and the values it reflects. In the case of climate change, however, the public interest objective appears to be straightforward: it is the reduction of carbon emissions, which many governments are now specifying in quantitative terms. It is true that much is contested behind these values, such as the science of climate change and what sacrifices should be made now for future benefit, but overall governmental objectives are clear and are not couched in general principles.

Climate change caused by greenhouse gas emissions is very often described as a classic case of market failure (Stern 2007), and the main policies and regulations, being designed to internalise the externalities, appear to be clearly explicable by public interest theory. The policy most explicitly focused on internalising the externality of greenhouse gas emissions is putting a price on carbon. This policy has been widely advocated. In his Nobel Prize acceptance speech in 2007, for example, Al Gore noted: 'and most important of all, we need to put a price on carbon' (Gore 2007). The cost to the environment is internalised by pricing carbon emissions at a level that forces the reductions necessary to avoid dangerous climate change. Carbon can be priced by a tax, which sets the price, or by a cap and trade system, which sets the quantity of emissions reductions. In Britain and Europe the EU's cap and trade scheme covers about 50% of greenhouse gas emissions. From a public interest perspective, cap and trade has the advantage of focusing on the principal public interest objective, namely capping and reducing the level of emissions. The carbon price is set in the market depending on the cost of reducing carbon and the demand for carbon-intensive products. Setting a carbon price is accompanied by a wide range of other policies, such as technological support and information and education. These developments indicate a policy approach moving towards a stronger and more explicit public interest position.

Failure to meet the public interest objective

Despite the climate change regulations in operation, evidence of success is limited. The first phase (2005–2007) of the EU's emissions trading scheme has been beset with problems. In particular, large industrial carbon emitters have been very concerned about the potential costs of the scheme, and emissions allowances were initially freely allocated on the basis of established emissions rather than being auctioned. Complex rules, with special allocations to certain interests, and over-allocation led to a big drop in the carbon price and significant financial windfalls to industry (Baldwin 2008, pp. 197–198). More established policies, such as those in the UK's Climate Change Programme, have also had limited success. The domestic targets set for 2010 in the 2000 programme will almost certainly not be met. In the 2006 programme, projections for emissions cuts by 2010 were significantly reduced compared to those in the 2000 programme – in 2000 the projection for CO_2 reductions by 2010 was 19% below the 1990 level while in 2006 it was only about 10% below (DETR 2000, DEFRA 2006). Uncertainties in emissions projections were recognised in the 2000 programme (DETR 2000, p. 127) and in 2006 the lower than expected reductions in CO_2 emissions were ascribed to higher than anticipated economic growth and changes in the relative price of coal and gas (DEFRA 2006, p. 3).

A particular problem is that, with the exception of the EU emissions trading scheme, quantified emissions reductions connected to the aggregate targets are not built into specific policies. Policies and regulations are dispersed and proliferating in an apparently ad hoc and disconnected fashion (Helm 2005, pp. 27–29). There is a complex landscape of policies covering a wide range of sectors and deploying a variety of regulatory tools, including direct regulations, trading schemes, taxes, voluntary agreements, information and education. Many policies appear to overlap with the EU emissions cap and trading scheme (Bartle and Vass 2007). For example, in the UK the EU scheme creates an incentive for the electricity generators to move to low carbon emission technologies. At the same time a very different policy, the Renewables Obligation, places an obligation on electricity generators to produce a pre-specified amount of renewable electricity. While the British government has recognised some overlap (DEFRA 2006, p. 49, 2007a), the general approach has been the development of disconnected policies on a sector by sector basis. The new Climate Change Act (2008) appears to be a recognition of the limitations of previous policies and is focused on the public interest objective through the use of increasingly tighter carbon 'budgets' (DEFRA 2007b). However, the processes of enforcement and compliance are untried and untested, and reductions remain dependent on the actual (and uncertain) performance of particular instruments and on factors such as economic performance, infrastructure investment and consumer behaviour.

Better regulation, transparency and the market failure approach

Although public interest theory appears to explain credibly climate change policy objectives set by government, it seems less able to explain the specific regulations and their implementation. The latter, notably the emissions trading scheme, appear to be better explained by the influence of private interests, particularly large polluting industries. This interpretation seems to offer a bleak prognosis regarding the need to shift towards more public interest orientated climate change regulation. At best it seems dependent on private interests gravitating towards the public interest. Although there is some evidence that this is beginning to happen – a major report from the Confederation of British Industry, for example, has called for a clear long-term regulatory framework for the reduction of greenhouse gas emissions (CBI 2007) – it is not clear that it will happen fast enough.

However, there are reasons not to be too pessimistic about the role of private interests in regulation. To assume that policy makers are driven primarily by sectional interests or self-seeking motives is arguably too cynical (Ogus 2004). The EU emissions trading scheme is being tightened and there is a move towards auctioning of allowances despite strong resistance from industry interests. In addition, regime analysis indicates that regulatory regimes are less static than private or public interest theories presume: they vary significantly across sectors and change over time (Hood *et al.* 2001). This variation suggests the possibility that governments can take action to change the balance of regime forces.

One such possible action to prompt change is a governmental drive towards transparency. Transparency is widely perceived to be a key means of improving modern government. In a volume which is particularly positive about the benefits of transparency, it is argued that 'transparency can be an effective, sometimes a transformative, tool serving the public interest' (Florini 2007, p. 14) and it is concluded that transparency improves the effectiveness and efficiency of modern government (Florini 2007, p. 338). Another contribution to the volume argues that transparency can also reduce the dominance of powerful interests in policy making and enhance trustworthiness and legitimacy (Ramkumar and Petkova 2007, p. 305). In a publication more critical of transparency it has been noted that there it can have an instrumental value, with synergies and trade-offs with other values such as efficiency and legitimacy (Heald 2006, pp. 59–73).

More particularly, in relation to regulatory regimes, 'openness and transparency come as close as any other contemporary doctrine to being an all-purpose remedy for misgovernment, because of its claimed ability to reduce corruption and transaction costs, increase legitimacy and legality, and improve policy quality through enhanced intelligence and learning' (Hood *et al.* 2001, p. 148). Transparency has also become a central theme in the 'better regulation' agenda pursued by governments since the 1990s (Hood *et al.* 2001, pp. 147–148) and is one of the UK government's five principles of 'good regulation'. Pressures for transparency are subject to 'institutional distortion' and changes are not necessarily always positive (Hood *et al.* 2001, pp. 168–170). Nevertheless, enhancing transparency appears to be a technique which might enable climate change regimes to move towards meeting the public interest objective.

Climate policy at the macro level in the UK has been fairly transparent, and there are prospects for improvement. The Climate Change Programme 2000 specified a target of reducing emissions by 20% of 1990 levels by 2010. The Climate Change Act appears to enhance transparency with annual reduction targets set within a series of five-year carbon 'budgets'. There is nevertheless scope for improvement.

It is, however, the micro level and its connection to the macro level which can particularly benefit from more transparency. The proliferation of micro policies and regulations means that few understand the policies and policy mix, and little academic work has been undertaken on their interaction (Boemare *et al.* 2003, Hasselknippe 2003, Sorrell and Sijm 2003, Sijm 2005). It is unclear how individual policies fit together to produce the required emissions reductions. In particular, there could be greater transparency about how the policies will contribute to the required quantified reductions and about what is, and what is not, quantitatively controlled.

The European Commission's proposals in 2008 do involve an increase in transparency (Commission 2008) in relation both to what is and what is not quantitatively controlled, and to what is controlled at the EU level and at national level. As such they represent a step towards greater transparency. All the EU-controlled emissions are to be directly controlled via the cap and trade scheme, while the instruments used to control the majority of national-level emissions will not be directly connected to a quantified cap. In the UK, for example, about 45% of emissions are covered by the EU trading scheme (35% energy supply and 10% large industry). The remaining 55% are not quantitatively controlled (21% transport, 14% domestic, and 20% small and medium businesses, public sector organisations and agriculture (2006 figures)) (DEFRA 2006, UNFCCC 2007).

Despite these positive steps and the stress on market failure and the carbon price approach, there is a lack of transparency about the two main policies to set a carbon price, namely tax and cap and trade. Cap and trade appears to be the main carbon price policy instrument. The EU's emissions trading scheme covers around 50% of emissions, and in the UK other cap and trade schemes have been proposed, such as the mandatory Carbon Reduction Commitment, which from April 2010 will provide a financial incentive to improve energy efficiency for over 5000 organisations not covered by the EU scheme or voluntary Climate Change Agreements. There remain some important taxes, notably fuel tax and the Climate Change Levy (an energy tax on industry), but the latter overlaps to some extent with the EU trading scheme and the Carbon Reduction Commitment, and might be reduced or phased out (DEFRA 2007a).

Despite cap and trade being the main policy approach, much political and media discourse is orientated towards 'green' or 'carbon' taxes. For example, a UK parliamentary report published in early 2008 notes that 'the government must take great care in designing environmental taxes, but nevertheless we are firmly of the view that environmental taxes are a useful and valuable tool to combat carbon emissions' (House of Commons 2008, p. 34). The report also offered cautious words in support of trading schemes, but provided little analysis of which would be appropriate in what circumstances. In addition, a press account of a report by the UN's Intergovernmental Panel on Climate Change (IPCC) was headlined 'UN scientists urge carbon tax', while noting that the IPCC does not explicitly recommend a tax but only an 'effective carbon price signal', and that 'Europe favours cap and trade systems' (*The Guardian*, 17 November 2007, p. 6). Without a clear statement of policy purpose, green taxes can appear to be just another way of raising revenue.

There is therefore scope for increased transparency about the kind of carbon price instrument adopted and its scope. Greater transparency with the carbon price instrument offers the possibility of positively influencing public opinion, thereby enabling the regulatory regime to move towards meeting the public interest. It might also lead to greater policy credibility and stability, and appeal to industry interests which in recent years have come to support a clear emissions reduction regime (CBI 2007).

Social plurality and the legitimacy problems of the market approach: the need for transparent core and complementary policies

The implications of the market failure approach using a transparent carbon price instrument, such as cap and trade or tax, are appealing. Micro policies and their implementation could be clearly connected to macro objectives, and clarity of policy purpose could offer the possibility of increased support, acceptance and political legitimacy. However, there are limits to what transparency can achieve. While the case for greater transparency in governance and regulation appears self-evident, it is not a panacea for all ills. In response to pressures for greater transparency, regulatory regimes can adapt negatively as well as positively (Hood *et al.* 2001, pp. 148–170), and mechanisms of transparency can be tokenistic and fail to address substantive problems (Hood 2006, pp. 223–224).

In particular, there might be substantive problems with the extension of the deployment of cap and trade, if it involves replacing existing policies, which increased transparency is not likely to be sufficient to address. Many existing policies are embedded, established and to some degree proven with stakeholder commitment. The effects of newly introduced policies are largely uncertain despite econometric analysis in documents such as the Stern report (Stern 2007), and a priori cost–benefit analysis of climate policy and regulation is imbued with uncertainty (van den Bergh 2004). Furthermore, the price signal on its own might lead to failure to invest in research and development and in infant technologies and industries.

There are also doubts about the fairness and legitimacy of emissions trading schemes (Baldwin 2008). In theory it is possible to separate questions about equity and efficiency in trading schemes, in so doing making such schemes more transparent and amenable to good decision making. The distribution of emissions allocations determines equity, and decisions could be made democratically while efficiency is achieved by the trading mechanism within the market. However, in practice this separation is difficult and the market mechanism often dominates and exaggerates inequalities by favouring the well-informed and well-resourced (Baldwin 2008, pp. 203–204). This is particularly the case in the EU scheme in which the emissions of large polluters are controlled. Allocations, which are a possible means of promoting equity, are made among large polluters, not among consumers or citizens.

More fundamentally, there are drawbacks with the market failure approach itself: as this defines people in terms of their economic interests, greater transparency would be politically effective only insofar as this market-based idea of interests is acceptable and legitimate.

More precisely, the market-based approach to the public interest and the deployment of the price instrument is founded on a single form of human behaviour: the rational economic actor – 'homo economicus' – who responds only in a self-interested way to price signals. However, there is a significant body of sociocultural research which suggests that the ways in which individuals and organisations think and act vary (Thompson *et al.* 1990). In particular, it is suggested that there are four contrasting sociocultural frames which condition people's thoughts and actions: egalitarianism, hierarchy, individualism and fatalism (Thompson *et al.* 1990, pp. 5–11). While it is acknowledged that individuals often inhabit more than one frame, and that there are 'multiple selves', it is argued that people often inhabit one frame more than others, and think and act differently according to the dominant frame (Thompson *et al.* 1990, pp. 265–267).

In this connection climate change is a particularly 'contested terrain', with distinct sociocultural variety (Verweij *et al.* 2006, p. 822). Egalitarians emphasise the damage caused by profligate patterns of production and consumption in developed countries. They stress that more equitable relation-ships are required between humans and between humanity and nature. Hierarchicalists stress that better national and global governance and planning is required to ensure that the natural world and its resources are better managed. Individualists are generally sceptical about global warming and the need for planned collective change, stressing that human beings are resourceful and innovative and will respond with technical developments driven primarily by market forces. Finally, fatalists see no reason in nature and little possibility of effecting change.

Sociocultural variety can create problems for many specific policy approaches, including price-based approaches. The dual approach of hierarchical imposition of a tax or cap combined with individualist market action might seem advantageous, but both hierarchalists and individuals may react adversely, hierarchicalists because they expect more direct action by government, such as regulations (directly curtailing polluting activity) to drive change, and individualists because they are sceptical about collective action against climate change. Duality can also add to complexity, which is a significant problem in cap and trade schemes, and can lead to vulnerability to industrial lobbying (Baldwin 2008, p. 204).

The price mechanism is also particularly problematic for egalitarians, as it aims to exploit the rational, self-interested, utility-maximising individual whereas for egalitarians it is precisely this individual, with their disconnection from nature, that is causing the problem. The price mechanism may or may not change contingent behaviour, but egalitarians want social and attitudinal change to be more deeply embedded. There is also a problem of social equity. If emissions caps become tight and carbon is scarce, its price – that is the price of heating, transport, food and other goods and services – will rise substantially and affect the poor disproportionately. Given the severity of climate change, egalitarians argue that there should be some equality in reducing carbon emissions, and that this would add to the legitimacy of climate policy.

A strategy for better climate change regulation: transparent core and complementary policies

This discussion suggests at least that multiple rather than single policy approaches should be developed. In this vein Verweij *et al.* (2006, p. 840) propose that 'clumsy' policies and institutions are required to ensure 'the effectiveness of attempts to tackle major social problems and the legitimacy of this process'.

> Clumsy solutions are creative, flexible mixes of four ways of organising, perceiving and justifying that satisfy the adherents to some ways of life more than other courses of actions, while leaving no actor worse off. As such, they alleviate social ills better than other courses of actions. (Verweij *et al.* 2006, p. 840)

Thus 'clumsy' solutions offer a happy medium between the 'monologue' process of 'single-metric rationality' and the 'incoherence of complete relativism' (Verweij *et al.* 2006, p. 839).

However, there is still a lack of certainty about whether a clumsy approach would achieve the specific emissions reductions required. A strategy for better climate change regulation requires more clarity and certainty while avoiding the problems of single policy approaches.

One such strategy could be greater transparency about policies and regulations which provide the most credible way of achieving quantified reductions, accompanied by other policies which exploit different ways of thinking and behaving: a 'core' policy with quantified reduction aims plus a series of 'complementary' policies that can reinforce the core policy.

Complementary policies can draw on a variety of sociocultural frames and suggest different regulatory approaches that can complement an 'individualist' or market-based approach focused on economic incentives and a carbon price.

In a hierarchical approach the state leads with command and control regulations and policies. Government led and funded science, technology and innovation policies aim to make low carbon technology available at lower costs. Infant industry policies focus on establishing the technologies in the market. While non-technology-specific policies, such as the EU's trading scheme, favour established and lower cost technologies, the renewables obligation could be set up to focus investment on less established technologies and thus act as an infant industry policy (Sorrell and Sijm 2003, p. 429). Reforms to the renewables obligation announced in the April 2009 Budget that discriminate in favour of offshore energy indicate that the government has at least partially adopted this approach. Many direct regulations are hierarchical, with the objective of prohibiting or controlling particular activities and actions. Product standards, for example building or consumer goods standards or prohibition of certain goods, can reduce the demand for energy while the prohibition or direct control of certain greenhouse gases, such as CFCs and methane, can reduce emissions. Infrastructure policies, including planning rules and processes associated with transport and energy networks and the built environment, are hierarchical and can also affect greenhouse gas emissions.

Egalitarian approaches based on information and education focus on attitudes and social norms in changing behaviour. 'Information asymmetry' prevents market-based mechanisms from operating effectively, from which it follows that better knowledge can improve the working of the price mechanism. If and when prices rise, increased consumer knowledge of alternative lower carbon options can reduce the demand for high carbon goods and services. Information and education can increase awareness and bring about attitudinal change, and can also increase the acceptability and legitimacy of carbon price approaches. Legislators might then have the confidence to introduce more stringent policies. Behavioural change thus becomes more deeply embedded: it is automatic and spontaneous, rather than being contingent on financial incentives.

Complementary policies can be consistent and mutually supporting. Innovation and technology policies, for example, can increase the supply of low carbon technology and reinforce the price signal. Targeted assistance for technologies in earlier phases of their product cycle can co-exist with market orientated policies for more established technologies (Midttun and Gautesen 2007). Consumers may or may not accept higher energy prices caused by higher carbon prices, but not be informed or motivated to pursue realistic options to reduce consumption. Education and information could therefore reduce the demand for high carbon products and services and again reinforce the price signal. Policies other than those based on price might address the target problem in a different way, such as by increasing the provision of low carbon technology or lowering consumer demand, and could reinforce the direct price mechanism. Recognition of these processes is crucial in order to identify whether policies are complementary, overlapping or contradictory.

A key government strategy therefore is to be explicit and transparent about the ways in which policies are complementary in order to avoid the appearance of policy proliferation. The purpose and approach of each instrument and their linkages to others, particularly price instruments, needs to be made transparent. There are some indications that the UK government is beginning to package and present policies in terms of their type and policy purpose rather than categorising them into sectors and sub-sectors. The Stern report suggested three broad policy approaches (a carbon price, technology and innovation, and removal of barriers to behavioural change such as information asymmetry) (Stern 2007, p. 349) which correlate with the core and complementary policies. In its 'analytical audit' of climate change policies, the Office of Climate Change has suggested classifying policies in this manner. However, other potentially positive initiatives, such as the 'Climate Change Simplification project' (DEFRA 2007a), have been more focused on reducing the 'regulatory burden' on business and industry.

Complementary policies and decentred regulation

The focus in this contribution has been on established public and private interest theories of regulation and on a narrow notion of regulation as a set of rules determined by political authority. In the last decade, developments in regulatory theory have moved towards a broader notion of regulation which includes wider forms of social and economic control by the state and by actors and institutions beyond the state. In this sense regulation is a more dispersed and plural phenomenon than that which emerges from authoritative state rules. This has been conceptualised as 'decentred regulation' (Black 2001), and implies significant limitations on what the 'post-regulatory state' can do (Scott 2004).

A strategy based on core and complementary policies appears to be compatible with decentred regulation at least to a degree, although full analysis is beyond the scope of this study. It is particularly the need to recognise and draw on social variability in climate change regulation which parallels the dispersed processes of control in decentred regulation. The idea of decentred regulation suggests that the proposal for better complementary policies means not only the need for complementarity between different types of state rules but also the need for complementarity between state regulations and non-state forms of control. As Scott (2004, p. 153) notes, 'the successful implementation of regulatory law is dependent on achieving some measure of "structural coupling"', that is, a coupling between legal and non-legal forms of social and economic control. Successful coupling will not be easy to achieve. As Baldwin (2008, pp. 208–209) notes in relation to trading schemes, the coupling of democratic and market processes is fraught with difficulties and can favour better resourced and informed actors. Nevertheless, if sociocultural plurality and multiple policy approaches are accepted, the effective coupling of different processes needs to be addressed by the state. Arguably the literature on decentred regulation underplays the continuing importance of the state, which remains crucial in ensuring that regulation reflects public purpose and values

(Vincent-Jones 2002). Although it must undertake a more nuanced strategy than traditional regulation would suggest, it is the state which, as the single most important player in regulation, can influence the coupled zones of decentred regulation.

Conclusion

Does the theory and analysis of regulation provide an understanding of climate change regulation that can lead to political strategies to improve policy and regulation? In the regulation literature, public and private interest theories have been predominant since the 1960s. Public interest theory of regulation is appealing because of its clear connection to regulatory purpose and the need to overcome the sorts of market failures that are so significant in the case of climate change. However, regulation in practice rarely conforms to the public interest ideal, an observation that the record on climate change regulation in developed countries appears to support. Private interest theories might better explain actual regulatory outcomes, but empirical evidence indicates that these too are limited in their explanatory power.

Regulatory regime analysis provides a way of understanding different dimensions and forces within regimes. Depending on circumstances and context, regimes can be formed by (at least) three different dynamics: public interest/market failure, private interest and public opinion (Hood *et al.* 2001). Regime theory suggests that regimes are not static and subject to one force. They can change and be changed by any number of alterations in context and circumstances, including inputs from government. There is a literature that suggests that greater transparency could promote the effectiveness and efficiency of government and empower ordinary people against powerful sectional interests (Florini 2007). This suggests that a drive for transparency by government can alter the balance of forces within regulatory regimes and orientate them more towards meeting public interest objectives.

In climate change, greater transparency is required in the way in which quantified reductions are to be achieved. A seductively appealing way to achieve this, and one that draws on the market failure strand of public interest theory, is to adopt a more transparent carbon price instrument. The price of carbon, set by either a tax or cap and trade scheme, could then be clearly connected to the aggregate emissions reductions requirement. Cap and trade has particular appeal because of its direct connection to quantified reductions. In the UK and the EU there are movements in this direction, with about half of greenhouse gas emissions covered by the EU's trading scheme. Moreover, in 2008 the Commission proposed linking the trading scheme emissions cap to long-term targets and setting binding national targets for the reductions of emissions not covered by the trading scheme. Despite this, transparency remains limited, especially in relation to how quantified reductions are to be achieved in those sectors controlled at national level, notably transport and domestic and small business heating, which together account for about 30% of emissions.

Although a single transparent policy instrument which is clearly connected to the aggregate targets is appealing, there are reasons to doubt its likely efficacy and legitimacy. As Verweij *et al.* (2006, p. 839) suggest, 'a single policy approach is seductively elegant and free from the defiling intrusion of politics', but is severely limited. Successful reorientation towards the public interest requires more than a transparent and 'rational' policy; workable, effective and legitimate policies are also needed. Sociocultural plurality suggests that these are not easily accomplished, and that multiple policy approaches or 'clumsy' policies are required (Verweij *et al.* 2006). While drawing on these ideas, there is a need to go beyond the uncertainty implied by a clumsy approach to suggest a range of 'complementary' policies to augment a 'core' price-based policy instrument. The intention is that this approach would not only reduce the burden on the core carbon price-based approach by addressing different behavioural processes but will also, by addressing multiple sociocultural frames, increase the possibility of achieving legitimacy. In this way the climate change regulatory regime can be reoriented towards meeting the public interest objective and be more likely to win broader support.

References

Baldwin, R., 2008. Regulation lite: the rise of emissions trading. *Regulation and Governance*, 2, 193–215.

Baldwin, R. and Cave, M., 1999. *Understanding regulation. Theory, strategy and practice*. Oxford: Oxford University Press.

Bartle, I. and Vass, P., 2007. Climate change policy and the regulatory state. A better regulation perspective, *CRI Research Report 19*, Centre for the Study of Regulated Industries, School of Management, University of Bath.

Black, J., 2001. Decentring regulation: understanding the role of regulation and self-regulation in a postregulatory world. *Current Legal Problems*, 54, 103–147.

Boemare, C., Quirion, P., and Sorrell, S., 2003. The evolution of emissions trading in the EU: tensions between national trading schemes and the proposed EU directive. *Climate Policy*, 3 (S2), S105–S124.

BRTF (Better Regulation Task Force), 2003. *Principles of good regulation*. London: Better Regulation Task Force.

CBI (Confederation of British Industry), 2007. *Climate change, everyone's business*. A report from the CBI climate change task force. London: Confederation of British Industry.

Commission, 2008. Communication from the Commission to the European Parliament, the Council, the European Economic and Social Committee and the Committee of the Regions. *20 20 by 2020, Europe's climate change opportunity*. COM(2008)30, Brussels, 23 January.

DEFRA (Department of Environment, Food and Rural Affairs), 2006. *Climate change. The UK Programme 2006*. Cm6764, March. London: Department of Environment, Food and Rural Affairs.

DEFRA, 2007a. *Consultations on the recommendations of the Climate Change Simplification Project. Climate change instruments, areas of overlap and options for simplification*. December. London: Department of Environment, Food and Rural Affairs.

DEFRA, 2007b. *Taking forward the UK Climate Change Bill: the government response to pre-legislative scrutiny and public consultation.* October. London: Department of Environment, Food and Rural Affairs.

DETR (Department of Environment, Transport and Regions), 2000. *Climate change, the UK programme.* Cm4913, November. London: Department of Environment, Transport and Regions.

Feintuck, M., 2004. *The public interest in regulation.* Oxford: Oxford University Press.

Florini, A., ed., 2007. *The right to know. Transparency for an open world.* New York and Chichester: Columbia University Press.

Gore, A., 2007. *Nobel Lecture, Oslo, Norway* [online]. Available from: http:// nobelprize.org/nobel_prizes/peace/laureates/2007/gore-lecture_en.html [Accessed 8 April 2009].

Hasselknippe, H., 2003. Systems for carbon trading: an overview. *Climate Policy,* 3 (S2), S42–S57.

Heald, D., 2006. Transparency as instrumental value. *In*: C. Hood and D. Heald, eds. *Transparency. The key to better governance?* Oxford: Oxford University Press, 59–74.

Helm, D., 2005. Climate change policy. A survey. *In*: D. Helm, ed. *Climate change policy.* Oxford: Oxford University Press, 11–29.

Hood, C., 2006. Beyond exchanging first principles. Some closing comments. *In*: C. Hood and D. Heald, eds. *Transparency. The key to better governance?* Oxford: Oxford University Press, 211–225.

Hood, C., Rothstein, H., and Baldwin, R., 2001. *The government of risk. Understanding risk regulation regimes.* Oxford: Oxford University Press.

House of Commons, 2008. *Climate change and the Stern Review. The implications for Treasury policy.* Treasury Committee, Fourth Report of Session 2007–08, January.

Jordana, J. and Levi-Faur, D., 2004. The politics of regulation in the age of governance. *In*: J. Jordana and D. Levi-Faur, eds. *The politics of regulation. Institutions and regulatory reforms for the age of governance.* Cheltenham: Elgar, 1–28.

Levine, M.E. and Forrence, J.L., 1990. Regulatory capture, public interest, and the public agenda: towards a synthesis. *Journal of Law, Economics and Organisation,* 6, 167–198.

Midttun, A. and Gautesen, K., 2007. Feed in or certificates, competition or complementarity? Combining a static efficiency and a dynamic innovation perspective on the greening of the energy industry. *Energy Policy,* 35, 1419–1422.

Ogus, A., 2004. W(h)ither the economic theory of regulation? What economic theory of regulation? *In*: J. Jordana and D. Levi-Faur, eds. *The politics of regulation. Institutions and regulatory reforms for the age of governance.* Cheltenham: Elgar, 31–44.

Ramkumar, V. and Petkova, E., 2007. Transparency and environmental governance. *In*: A. Florini, ed. *The right to know. Transparency for an open world.* New York and Chichester: Columbia University Press, 279–308.

Scott, C., 2004. Regulation in the age of governance: the rise of the post-regulatory state. *In*: J. Jordana and D. Levi-Faur, eds. *The politics of regulation. Institutions and regulatory reforms for the age of governance.* Cheltenham: Elgar, 145–174.

Sijm, J., 2005. The interaction between the EU Emissions Trading Scheme and national energy policies. *Climate Policy,* 5, 79–96.

Sorrell, S. and Sijm, J., 2003. Carbon trading in the policy mix. *Oxford Review of Economic Policy,* 19 (3), 420–437.

Stern, N., 2007. *The economics of climate change, the Stern Review.* Cambridge: Cambridge University Press.

Stigler, G.J., 1971. The theory of economic regulation. *Bell Journal of Economics and Management Science,* 2 (1), 3–21.

Thompson, M., Ellis, R., and Wildavsky, A., 1990. *Cultural theory.* Boulder, CO and Oxford: Westview Press.

UNFCCC (United Nations Framework Convention on Climate Change), 2007. *Emissions summary for the United Kingdom of Great Britain and Northern Ireland (1990–2006)* [online]. Available from: http://unfccc.int/ghg_emissions_data/ghg_data_from_unfccc/ghg_profiles/items/3954.php [Accessed 26 February 2008].

van den Bergh, J.C.J.M., 2004. Optimal climate policy is a utopia: from quantitative to qualitative cost–benefit analysis. *Ecological Economics*, 48, 385–393.

Verweij, M., *et al.*, 2006. Clumsy solutions for a complex world: the case of climate change. *Public Administration*, 84 (4), 817–843.

Vincent-Jones, P., 2002. Values and purpose in government: central–local relations in regulatory perspective. *Journal of Law and Society*, 29 (1), 27–55.

Viscusi, W.K., Harrington, J.E., and Vernon, J.M., 1995. *Economics of regulation and antitrust*. 2nd ed. Cambridge, MA and London: MIT Press.

The (non-)politics of managing low carbon socio-technical transitions

Ivan Scrase and Adrian Smith

Whilst not originating in political analysis, a transitions problem framing nevertheless provides a heuristic for understanding certain features of climate politics. A multi-level perspective on transitions to low carbon socio-technical regimes is introduced, and illustrated through the example of electricity supply. An associated 'transition management' (TM) approach, which considers policy measures to induce low carbon transitions, is then adopted as the central focus. A diagnosis of the climate challenge from a TM perspective is developed, and a range of familiar strategies for reducing emissions is considered in light thereof. Although political realities and difficulties have been given little emphasis in the TM literature, strategies with political implications are identifiable. These recommendations have technocratic overtones and assume a consensual or corporatist model of politics. A committed and explicitly political programme for transitions, rather than TM, may be necessary.

Introduction

An enduring faith in technical fixes to environment–society problems is common amongst political leaders. In September 2007 President George Bush declared that greater use of carbon capture and storage (CCS), nuclear power, biofuels and other technologies will overcome major economies' climate and energy security challenges:

> We've identified a problem, let's go solve it together. We will harness the power of technology. There is a way forward that will enable us to grow our economies and protect the environment, and that's called technology. We'll meet our energy needs. We'll be good stewards of this environment. Achieving these goals will

require a sustained effort over many decades. This problem isn't going to be solved overnight. (Bush 2007)

Despite the caveats, Bush essentially argues that technology will solve society's problems without any need for significant social change. The UK government, by contrast, has sometimes acknowledged the important social dimensions associated with technological strategies:

> A low carbon economy is not a slogan. It will entail, over the next few decades, the transformation of our lives and of our economy – as the Prime Minister has put it, a 'technological revolution' in the way we use and source our energy. And in turn – because energy use pervades every aspect of our lives – this will imply a social transformation, in the way we live. (HM Government 2008, p. 2)

However, it remains a moot point in policy debates whether the low carbon technology comes first, inducing wider social changes (as the quotation implies), or whether social changes are required first in order that low carbon technologies become valued and developed.

The socio-technical transitions literature argues that social change and technology development are inseparable and have to be analysed accordingly. The literature exhibits a twofold purpose.

Firstly, it sets out to explain major shifts in 'socio-technical regimes' and attendant social transformations, such as 'electrification in western society' (Hughes 1983). Analysis draws upon theoretical resources from across the social sciences, particularly innovation studies, sociology and institutional theory (for example, Geels 2004) and, latterly, political science and theories of governance (for example, Meadowcroft 2005, Smith *et al.* 2005). This literature has found that major transformations are realised over decades through (unpredictable and often disruptive) evolutionary social and technological change.

Secondly, a policy-oriented part of this literature, known as 'transition management' (TM), argues for the deliberate realisation of sustainability-oriented visions of the future with technologies at their focal points. In this view, low carbon transitions can be achieved through consensus on visions and through common effort under a process-oriented, interactive style of national policy-making, implementation and learning. This second purpose is more explicitly normative (Elzen *et al.* 2004, Weber and Hemmelskamp 2005). TM seeks to use analytical lessons to guide purposive efforts in low carbon transitions where existing regimes are identified as unsustainable. TM's policy prescriptions (Rotmans *et al.* 2001, Kemp and Loorbach 2003, Loorbach, 2007) originate in Dutch researchers' advice to policy-makers, although its application by the Dutch government since 2001 has not fully matched researchers' aspirations (Kemp *et al.* 2007).

Sympathetic critics (e.g. Berkhout *et al.* 2004, Shove and Walker 2007) note that while past transitions were unplanned, emergent phenomena, TM seeks to steer evolutionary dynamics towards specific visions. Moreover, identifying the boundaries of a specific 'regime', reaching consensus over what needs changing, developing visions for future development and so on are extremely difficult not only in analytical terms but also politically, once real steps are needed to promote change. Problem framing is the critical point of departure for TM, but the possibilities and consequences of frame conflicts for TM are only just beginning to be considered.

The analysis begins by introducing the socio-technical transitions perspective in more depth, using the electricity regime as an illustration. The following sections sketch how a TM perspective diagnoses and prescribes policy measures for a number of specific carbon mitigation strategies. In the fourth section we consider a series of political strategies in relation to TM prescriptions ranging from tactical improvements to the basic TM approach to a more challenging consideration of the politics of low carbon transitions. Finally, we conclude that transitions need a broader political programme than TM to underpin them.

A socio-technical analysis of the climate challenge

The literature on 'technologies in transition' (Shove and Walker 2007) is wide ranging and spans several decades. Although definitive synopsis is beyond the scope of this study, in this section we briefly present a recent multi-level perspective on socio-technical transitions.

The socio-technical transitions perspective provides a problem framing in which technologies are understood as just one element in the 'regimes' that provide for societies' needs. These regimes consist of stable but continually evolving configurations of artefacts, actors and institutions. Like Kuhn's scientific paradigms, they go through long periods of 'normal' change, but eventually cease to serve their purpose and become susceptible to replacement by another regime. Newly emerging regimes grow from successful application and learning in niche uses or 'protected spaces' in which the central technologies, which at this point are imperfectly developed and problematic against established criteria, are not exposed to the full selective pressures operating in the incumbent regime. Over time, and with support, successful niches permit innovations in their socio-technical configuration that enable them to attract further technical, economic and political commitment. Wind energy, solar energy, wave energy and others can be considered niche technologies in most OECD countries, since they neither attract the material and institutional support enjoyed by the incumbent regime nor benefit from equivalent 'protections' offered by public policy support.

Whilst niches provide a 'micro' level source of transformative agency, this influence is structured by incumbent regimes at the 'meso' level. Beyond this, it is the 'macro' level of 'socio-technical landscapes' that provides the broader structural context for niche–regime interactions. This 'landscape' consists of material, demographic, ideological and cultural processes that operate beyond the direct influence of actors in any given regime and provide 'gradients for action' (Geels 2004). Landscape processes, political or otherwise, bear down upon regimes, generating stress and creating opportunities. Broad societal concerns over climate change (some rendered collectively effective through emissions policies) are one such stress. The hope is that these processes of niche development and regime destabilisation will, over time, generate a transition to a low carbon socio-technical regime. This could involve a gradual evolution or a more disruptive transformation: the socio-technical transitions perspective provides a conceptual framework for considering both.

In typical affluent countries, electricity supply is organised around large, fossil-fuelled generating technologies feeding power into a centrally managed grid system. Its operation is assured through the investments of large utility companies overseen by independent regulatory agencies. Millions of end users are accustomed to services powered by electricity. These key actors operate within an energy policy framework that seeks three goals: secure energy supplies, economic efficiency (competition and low energy prices) and environmental acceptability (mainly lower greenhouse gas emissions).

Conceptualised as an evolving socio-technical regime, the electricity system is understood to be reproduced through relations among complex, shifting networks of actors, institutions (rules) and artefacts. Policy measures and market mechanisms are important for the development of rules and standard practices within these networks, but the capacity of new institutions and policy measures to effect change is strongly constrained by existing actors' commitments and by material infrastructures. Outcomes are unpredictable, as regimes and levels of governance interpenetrate one another and throw up and/or respond to unforeseen contingencies. What this means is that the 'right' policies may yet be inadequate to bring about purposive change in the face of powerful actors contesting future directions and competing over material commitments (Smith *et al.* 2005). Socio-technical regimes embody historically accumulated debates and commitments about the mainstream realisation of social needs. The centralised, fossil-fuelled electricity regime is obviously more established than alternative electricity systems even if climate concerns are eroding its social legitimacy. Debates over alternatives involve, in part, contests over desired forms of a future low carbon society, and are therefore inherently political.

Within a stable socio-technical regime the rules of the game are relatively clear to all those involved in its reproduction (businesses, policy-makers, research institutes, regulators, investors and end-users). Expectations about future trajectories of development are structured accordingly, with some discourses and initiatives appearing more credible than others according to the extent to which they align with the existing regime and the nature of regime members who share that expectation and/or can be enrolled to work towards it.

Actors benefiting from the status quo tend to lend their agency to defending and extending regimes through incremental innovations. Over long periods of time the 'endogenous renewal' of incumbent regimes can deliver significant improvements in environmental performance even if the basic principles, business models and end-user practices are unaltered (Smith *et al.* 2005). The accumulated unit efficiencies associated with the scaling up of thermal capacity in coal-fired and gas-fired electricity power stations is illustrative, even though emissions are still rising in almost all wealthy countries due to the fact that absolute growth in demand for electricity is overwhelming the effects of relative efficiency improvements in supply.

The incremental development and spread of a fossil-fuelled electricity regime facilitated the emergence of synergistic interdependencies that reach right through the economy and society (Hughes 1983). Incremental development in this socio-technical regime has provided almost universal access to cheap, reliable electricity. In combination with other socio-technical innovations (such as those in computing and communication), this opened profound possibilities for economic growth. More prosaically, lighting a room at night became no more than a matter of flicking a switch.

It is therefore unsurprising that considerable effort is put into the maintenance of these regimes. The biggest shift in electricity governance in wealthy countries since the creation of national infrastructures has been a process of privatising key assets and liberalising energy markets. In principle, liberalisation opens space for alternatives (Markard and Truffer 2006), but in practice the new institutional context applies criteria and imposes requirements that still favour established players, non-disruptive technologies and existing patterns of material infrastructure. In the UK, 'sweating existing assets' inherited at privatisation has been the predominant business activity in recent decades (Helm 2005). Globally, coal- and gas-fired generation technologies, which slot relatively easily into the liberalised institutional regime, are the favoured investment options.

As a consequence it is difficult to promote the evolution of radically different socio-technical regimes. The technological, institutional and political alignments that lock nations into certain forms of electricity provision are multiple, mutually reinforcing and complex. However, governments are coming to appreciate that cutting carbon emissions has to be treated as an urgent priority, and the policy landscape is changing.

In this context it is understandable that nuclear power and CCS are very appealing for many regime members, despite high associated risks and costs, since they promise to cut emissions without disrupting too many alignments and linkages in the existing socio-technical regime.

Alternatively, more radical changes may be needed, for example to more decentralised regimes based principally on energy efficiency and renewable energy. Transitions between radically different socio-technical regimes such as these have occurred in the past, and some of the conditions for such a transformation in electricity regimes appear to be emerging today. The alignments that make a regime appear robust in one period can render it brittle if they inhibit adaptation to wider changes over time. Climate change, coupled to fears over energy security, forces a re-interpretation of the effectiveness of energy-related regimes and erodes their legitimacy. Growing pressure opens opportunities for transformation.

Transition management's diagnosis and prescriptions

The socio-technical transitions perspective does not point to any easy answers as to what governments should do in these circumstances. However, a set of prescriptions derived from the TM literature does offer them a 'multi-level approach' derived from the 'niche–regime-landscape' conceptual framework presented above, which acknowledges many of the complexities and difficulties that governments will face in promoting low carbon transitions.

Whilst attention is directed mainly at the creation and nurturing of socio-technical niches, it is recognised that success also requires pressure to be brought to bear directly upon incumbent regimes (Kemp *et al.* 1998, Rip and Kemp 1998). What this means is that TM provides a framework for considering portfolios of policy measures that nurture low carbon niche developments, putting pressure on emissions from incumbent regimes and facilitating processes for niches to inform regime transformation. In so doing it draws attention to the complex socio-technical inter-relationships involved, the long time frames, the broad participation involving producers and consumers, the importance of social expectations, the value of nurturing niche technologies, and the significance of learning and reflexivity. Consequently, it provides a guide for facilitating and co-ordinating distributed actions, and can help diagnose where governments are pursuing strategies that are simplistic or that evidence suggests stand little chance of success.

Diagnosis of the politics of climate policy

TM considers climate politics proceeding from 'landscape-level' societal awareness of climate change to constitute a threat to the existing energy regime, which generates social and policy pressure for carbon emission reductions and becomes a source of stress that plays out in various ways across different regimes.

Pressure to cut emissions prompts incremental repair work within the existing regime but also brings in new performance criteria and unsettles institutions. This creates opportunities for the emergence of alternative niches capable of presenting compelling solutions to the carbon problem. Niche proponents then have to do considerable technical, economic and political work in order to improve niche performance against the reconfiguring mix of performance criteria and persuade others to move commitments from the existing regime into their niche. Competing niche actors pursue similar strategies. With appropriate support, it is argued, it is possible to guide the emergence of a low carbon regime from this multi-level dynamic. As mentioned above, however, past transitions have taken decades.

Furnished with this analysis, TM advances its model for policy-making. Climate policy, according to this perspective, is about understanding, harnessing and modulating these multi-level dynamics for particular outcomes. The model begins with the low carbon goal, which might be a national or sector emissions reduction target. A series of multi-participant 'transition arenas' are then convened, usually by government. Participants deliberate on those regimes in which emissions are most significant or potential for change is identified.

Each transition arena goes through an iterative process of: (i) understanding the nature of the carbon reduction challenge for the focal regime and identifying 'transition goals' such as 'to make electricity supply more sustainable'; (ii) developing a consensus about alternatives and a basket of 'visions' that is compatible with the transition goals; (iii) identifying 'pathways' towards those visions, such as expansion of energy generation from renewable sources; (iv) instigating niche experiments that contribute to the realisation of those pathways; and (v) establishing processes for social learning and reflexivity across all of these activities. Operating in a recursive cycle, transition policy measures build momentum behind successful niches and lead to a gradual transfer of institutional support away from the regime and towards a nascent low carbon regime emerging out of niches.

TM thus offers a nuanced and logical diagnosis of the politics of climate change, plus a set of policy prescriptions. Problem definition leads to solutions which guide activities, while outcomes inform revisions to the original problem definition, and so on around an iterative, reflexive and consensus-seeking cycle.

However, both diagnosis and prescriptions are problematic. While 'visions' and 'pathways' are open to wide participation *in principle*, for example, and open to revision as time goes on, the TM literature recommends that transition goals are agreed by 'a relatively small network of innovators and strategic thinkers from different backgrounds that discusses the transition problem integrally' (Kemp and Loorbach 2003, p. 16). This idea of small networks of actors ('transition arenas') driving agendas suggests a potentially elitist approach. Indeed, 'distributed control' is identified by Kemp *et al.* (2007) as a 'key problem' for TM. While 'allowing local government an amount of discretion regarding what to do' is recommended once local authorities have been recruited to the transition agenda, for instance, Kemp *et al.* (2007, p. 318) also recommend 'functional coordination, cross-linking and integration' in government. Emphasis on foresight and use of scenarios, back-casting, 'integrated appraisal' and related procedural tools enmeshes TM in a technocratic discourse, although tempered by appeals to reflexivity.

How then does TM approach the carbon abatement policy measures that are the focus of this volume? In the remainder of this section we provide a TM perspective on cutting transport-related emissions, carbon trading and energy taxes, tightening energy standards, expansion of renewable energy and nuclear power, and low carbon innovation policy.

Cutting transport-related GHG emissions

The TM approach seeks to assemble policy portfolios that contribute to the core transition processes. Policy measures aimed at cutting transport-related GHGs would be organised accordingly so that planning, regulation and market mechanisms are deployed in such a way as to improve the chances that sustainable niche technologies will displace existing transport regimes (Hoogma *et al.* 2002).

The starting point is problem structuring and goal envisioning. This might proceed as follows. First, the national transport department facilitates a multi-stakeholder forum, bringing in 'strategic thinkers' from vehicle producers, fuel suppliers, planners, consumer groups, environmental NGOs and so on. This 'transition arena' deliberates on the problems of the existing transport regime and the goals that ensuing transition initiatives must serve. Great emphasis is placed on mutual learning, consensus building and developing a shared problem perception (van de Brugge and van Raak 2007). Techniques such as foresight studies and scenario building are used.

A 'basket' of technology-centred sustainable transport 'visions' is then developed, such as 'zero carbon vehicles', 'ICTs minimising the need to travel' and 'trains and coaches as first choices for mobility'. Transition pathways and intermediate goals are developed using techniques such as back-casting (Quist 2007), and kept under review as events unfold. A portfolio of 'niche experiments' is created to explore different possibilities for initial steps along one or more potential pathways. Such experiments might involve electric vehicles, car clubs, multi-modal transport integration systems, or novel vehicle fuels. Promising socio-technical niches would then go through phases of 'pre-development', 'take-off' and 'acceleration', culminating in 'stabilisation' with a more environmentally benign regime (Rotmans *et al.* 2001). Some niches will fail and support will be dropped, while new niches will be continually brought into an evolving portfolio of activities. Outcomes are emergent, with surprises and disappointments, which underscores the need for adaptive governance.

'Learning' and 'adaptation' are the essential links between long-term goals, socio-technical pathways and short-term actions in niche experiments. Lessons are drawn not just for the niche itself but also for revising policies, visions and pathways in the light of actual progress towards the transition goals (Hoogma *et al.* 2002). Recent sustainability debates over biofuels, for instance, would inform a revision to policy support for this niche and direct more carefully any continued development.

The institutionalisation of niche-derived lessons is given relatively little attention in the TM literature even though it is recognised as being a vital element (Berkhout *et al.* 2004, Kemp *et al.* 2007). This is the most difficult step politically and economically as it is the moment when the lessons of niche experiments are translated into momentum for change. At this point serious commitments are needed in areas such as investment in infrastructure, altering market incentives, policy reforms, creating or reforming institutions, and confronting users' expectations.

Deeper institutionalisation of promising, alternative socio-technical configurations implies selection pressures becoming mobilised against the incumbent regime, which means that powerful actors' positions will be affected (Smith *et al.* 2005, Shove and Walker 2007). However, in a typical affluent country it is political arenas rather than transition arenas that are likely to become the determining sites of deliberation. The problem is that TM offers little guidance on political strategy once the potential gains from transition arenas are exhausted.

Carbon trading and energy/carbon taxes

Carbon trading and energy/carbon taxes provide landscape pressures essential for TM's aims. Although taxes are relatively inflexible once established, they would be a useful device for opening space for low carbon niches by making the 'gradients for action' more favourable to low carbon socio-technical practices and making carbon-intensive practices less profitable.

Political barriers, however, blocked a carbon tax at the EU level and in many member states, and the EU is strongly committed to carbon trading as the principal mechanism for promoting low carbon transitions. Carbon trading has the advantage of permitting much greater flexibility as to how required emissions reductions are achieved. The EU policy of gradually tightening up the caps, reducing national quotas, and auctioning allowances (as planned from 2013) fit well with TM's gradualist approach.

Both measures draw upon neo-classical economic theory, which presumes that technologies emerge in response to price signals. This perspective is illustrated in a letter in 2003 from Göran Persson and Tony Blair (the Swedish and UK prime ministers respectively) to Romano Prodi, President of the European Commission.

> New technologies and processes can contribute to the goal of decoupling economic growth from environmental degradation. However, to a large extent the technologies and systems of the past still dominate in important areas such as transport, energy, industry and agriculture. In each of these sectors new and better technologies are available or emerging. To speed up the replacement of old technologies there is a need to set clear targets, develop stronger market based incentives and make more use of the instrument of public procurement. (Persson and Blair 2003)

This view implies that better technologies are ready and waiting, in relatively autonomous forms, and that innovation and diffusion will happen with the right price signals and a little 'demand pull'. TM, however, draws on the perspective of evolutionary economics, which considers technology development to be highly uncertain, constrained by routines, and therefore sticky in relation to prices (Dosi 1982). Technology development is seen as an intrinsically social construction of the mutual alignments between artefacts, actors and institutions. Technologies have to be *made* to happen: while TM recognises the value of price signals, it argues that these alone are insufficient. In particular, the TM perspective reminds us that putting pressure on regimes without attending to niche solutions may simply result in additional support for technologies, such as CCS and nuclear, that are relatively compatible with existing regimes.

Tightening of energy standards

TM considers tightening energy standards to be a rather modest goal in itself. Rather, the measure should be formulated and implemented in relation to a systems transformation goal, such as 'carbon neutral buildings', and a vision/ pathway such as 'zero-carbon housing refurbishment'.

As an example, the housing sector in the UK is traditionally conservative towards innovations in general and low energy/carbon designs in particular. Improvements progress slowly through changes in building regulations. These standards are negotiated on the basis of incremental improvements deemed reasonable in the light of the existing housing regime, rather than negotiated on the basis of encouraging developments towards a vision for low carbon housing (Raman and Shove 2000).

TM, in contrast, looks for standards in forms that are innovation forcing and that capture benefits for first- and second-movers. Thus a TM approach would not develop standards through exclusive abstraction of technological and economic performance of elements of a house, but would instead proceed by analysing the interlinked socio-technical processes that constitute a low carbon house (Shove 1998, Rohracher 2001). Standards would be considered in the light of practices in existing low carbon housing niches and would point volume house-builders (and conventional householders) towards adopting practices found in those niches.

Standards in housing, as in other sectors, would also contribute to translations from niches to mainstream (Smith 2007). Understanding the housing socio-technical regime reveals, for example, that where new build rates are slow, retrofitting low carbon measures into the existing housing stock becomes essential. This implies that a portfolio of complementary policy measures is needed to help 'translate' niche practice into the mainstream norm, such as providing training in home improvement skills and help for householders to adopt new routines.

Expansion of renewable energy and/or nuclear power

Greater use of renewable energy is central to one of the 'transition visions' identified by Kemp and Loorbach (2003) and used in Dutch policy. Similar processes to those described for transport are needed here, but the focus in this subsection is on the implications for 'top down' processes that put pressure on the incumbent fossil energy regimes. This is recognised in TM as a necessary complement to stimulating niche technologies, yet in Dutch policy practice it is rather neglected (Smith and Kern 2009).

The European Union has set a target for 20% of its energy to come from renewable sources by 2020. Much greater use of renewable electricity (and/or nuclear power and CCS) is therefore emerging as an essential objective for national climate policies. Some renewables, like offshore wind, can be operated on such a large scale that they are effectively centralised sources that can feed into national transmission networks. However, experience in Denmark, Spain and elsewhere suggests that greater use of distributed renewables (connected to local networks) will also be needed.

Giving priority access to renewables, along with price support, has been central to Denmark's and Spain's success in increasing renewable energy generation. The consequent growth in renewables has required subsequent innovations in load management, network infrastructure investment, and related regulations. This experience is suggestive of how TM would operate: niches would be encouraged and promoted, and facilitating institutional reforms and material investments would be made to incumbent infrastructures on the basis of niche experience.

The UK government, in contrast, has decided that nuclear power and new coal-fired electricity generation are urgently needed for security of supply reasons (see *Guardian* 2008). However, the UK also faces a ten-fold expansion in renewable energy to meet its requirements under the EU Renewables Directive. The problem here relates to the extent to which the institutional and infrastructural pathways towards renewables options are compatible.

Given that nuclear power is low carbon, and that the industry has developed safer reactor designs, one might expect some sympathy for nuclear power amongst adherents of TM. However, TM's emphasis on bottom-up technical change and learning in niches, and its self-conscious positioning as distinct from blueprints and planning, means that this sort of technical fix appears ill-suited. Furthermore, given the large-scale, long-lived infrastructure and intense political commitments attending nuclear power, TM would caution against expanding its use because of concerns over technological 'lock-in' (Unruh 2002) and institutional 'entrapment' (Walker 2000).

The major implication of this diagnosis relates to system-level visions and priorities: if renewables are to be expanded, a more decentralised system appears appropriate and commitments to a centralised incumbent regime become problematic (Mitchell 2007). This involves making very hard political choices about institutions and infrastructures in contexts in which incremental improvements to the existing regime appear more favoured (as in the UK strategy) than the more uncertain process of transforming them.

Other policy measures emphasised in TM

The TM approach also emphasises other policy measures. These include innovation policies, measures to create transition arenas, and co-ordination measures for learning and action across policy portfolios.

Innovation policy, for example, traditionally focuses on promoting technologies that contribute towards competitive advantage and economic growth. This has conventionally been considered a linear process that proceeds through research, design, development, deployment and diffusion. Innovation policy supports the knowledge base, provides grants for development, assists with demonstration programmes, and encourages markets through fiscal incentives, public procurement and information services. More recently, innovation policy has recognised non-technological innovation, such as innovation in service provision. It has also adopted a less linear and more systematic appreciation of the socio-technical processes constituting innovations (OECD 1999). In some cases, innovation policy is also being targeted at social goals such as low carbon innovation (HM Government 2008).

TM puts innovation centre stage in its approach, and so places great emphasis on low carbon innovation policy. Policy measures which look at innovation through the wider lens of niche–regime interactions are particularly favoured, especially if coupled with the ambition of inducing a decarbonising transformation of incumbent, carbon-intense socio-technical regimes.

TM needs specific policy measures to create transition arenas. Superficially, this involves making life easier for transition stakeholders. More significant measures would then be needed to imbue the transition arena with legitimate authority sufficient to drive through their goals. This includes the ability to convey their consensus visions and pathways across government and its policy networks, the resources to support niche experimentation, and a capacity to require emission reductions in ways informed by TM.

Transition management's implications for political strategy

Advocates of TM argue that 'business as usual' policy-making is inadequate because it fails to address the system-level causes of climate change. System-level problems require 'a different policy approach: a long-term, integrated approach addressing problems of uncertainty, complexity, and interdependence' (Kemp and Loorbach 2003, p. 3). Strong policy co-ordination is required that integrates innovation, economic, environmental and other policy domains and institutionalises learning processes beyond electoral cycles and other political vicissitudes. In the Netherlands this co-ordination has been provided by the environment and economic affairs ministries through TM revisions to the National Environmental Policy Plans. However, there is little sense that achieving transitions is expected to become a political project as such.

When considering political strategies, one must consider what TM suggests, what can be revised (given early experience), and what non-TM strategies might be appropriate for transitions.

Political strategies identified or implicit in TM

Technocratic overtones in TM are tempered by humility in its analysis and prescriptions that reflects an appreciation of past transitions and insight across the social sciences. There remains, however, a sense that a total, 'integrated' package of policy-orchestrated low carbon transition is possible, and with it a frustration with politics caused by paying insufficient attention to the contested meanings of, and responses to, climate change. TM's principal architects position the approach as a new, evolutionary model of governance (Meadowcroft 2007). Planning and market mechanisms may be employed, but TM is seen as an overarching alternative to both.

The relatively late interest in politics among TM advocates may relate, in part, to its origins as a policy-oriented realm of academic thought developed in consultation with policy-making elites. Any political challenges inherent in TM prescriptions had to be downplayed in order to gain assent from policy elites (Smith and Kern 2009). Thus TM ideas are presented, as the name suggests, as managerial governance rather than politics (Smith and Stirling 2007). Some political tactics can be discerned, but political strategy is erased from the discourse.

An *implicit* political model is revealed, however, in the assumed relationship between transition 'goals' and their achievement. Once agreed, transition goals structure other efforts and must therefore be understood as part of the policy 'landscape'. This level is understood to be beyond the reach of actors at the level of any particular regime but is, by implication, instantly changeable by policy-makers, who are to be informed by a vanguard of transition managers acting on society's behalf. Popular support for these new goals is assumed to be possible to obtain.

The model assumes that true social values can be identified and distilled in this way, that policy emanates from values, that policy determines outcomes, and that social values can be decided upon, institutionalised and translated into policy action by policy-makers. Implicit in TM is a rationalistic model of policy in which niche lessons will be taken up and acted upon consensually. A more argumentative model would see those lessons being variously taken up, rejected or ignored by different actors, and debated and reshaped to suit their (re-considered) material and economic interests.

TM's only political strategies, as such, are to frame questions in an inclusive way and then to defer taking politically contentious actions. TM recommends that policy-makers should not define the details of change too early, on the grounds that one cannot know how events will unfold. By defining broad goals and visions, promoting experimentation and 'modulating' pressures for socio-technical change, it is hoped that a broad coalition can be constructed.

Ensuring that transition arenas have broad membership (even if they remain small groups) and that they develop multiple 'visions' (rather than a single 'blueprint') are political tactics of a kind (Meadowcroft 2007), but perhaps not ones well suited to politics as we know it. Even in the Netherlands, with its tradition of consensual politics, powerful groups threaten to co-opt and undermine the normative ideals of TM (Smith and Kern 2009).

TM tactics work only as long as real commitments are deferred. But one cannot avoid a hard look at the structural privileges of regime incumbents forever. Greater attention needs to be paid to preparing for 'the acute political struggles' to come (Meadowcroft 2007, p. 8). Perhaps this asks too much of TM, however. Kemp and Loorbach (2003, p. 12) stress that TM is 'a form of process management against a set of goals chosen by society. Societies' problem-solving capabilities are mobilised and translated into a transition programme, which is *legitimised through the political process*' (emphasis added). The ideas are sold in a way which appeals to the 'policy-makers' tasked with implementing the decisions of 'political actors': it appeals most, one suspects, to those in politics who see hope for a rational, orderly harnessing of technology for society's benefit. Keeping faithful to this aim suggests either that better TM is needed, or that the political context for TM needs to change.

TM strategies to overcome 'problems of steering'

Kemp *et al.* (2007) propose adjustments to TM that address five steering problems: 'dissent and ambivalence about goals', 'dealing with uncertainty', 'distributed control', 'political myopia', 'determination of short steps for long term change' and the 'danger of lock-in'. In so doing they reposition TM as a way to combine 'the capacity to adapt to change with a capacity to shape change' (Kemp *et al.* 2007, p. 316). However, they do not consider the possibility that TM itself might generate some of these 'problems', nor that dealing with them might be beyond its remit. For example, Kemp *et al.* (2007, p. 316) argue that a 'proximate solution to the problem of dissent is: continuous and iterative deliberation and assessment' using 'problem structuring methods' derived from complex systems theory. At the same time, however, TM is also creating a new process through which large numbers of potential entrepreneurs can access funds for niches, and therefore risks a descent into pork-barrel politics. Nevertheless appraisal methods can be useful in this situation. Cost–benefit analysis (CBA), for instance, originated in US politics relating to flood alleviation and river engineering and its impacts on land users and transport operators: it was an attempt to replace pork-barrel politics with 'mechanical objectivity' (Porter 1995). Although CBA can be criticised for closing down debate in certain ways, it is possible to use other forms of appraisal that genuinely open up and facilitate more transparent political debate (Stirling 2008).

Kemp *et al.* (2007, p. 317) recommend that risk assessments, technology assessments and monitoring be used to deal with the problem of long-term uncertainty: uncertainty about society's future needs and preferences can be partly dealt with by 'developing anticipatory strategies that help influence and change current preferences'.

Proceeding in this way favours experts who are able to operate within these kinds of appraisal discourses. Kemp *et al.* (2007, p. 326) express disappointment that Dutch TM practice has been 'dominated by regime actors', with 'outsiders ... scarcely involved', and note that the largest national consumer organisation had not participated because it felt it lacked relevant knowledge. They also note that transition arenas in the Netherlands have been dominated by representatives of business and academia. Smith and Kemp (2009) found NGOs opting out due to dissatisfaction with the process and its (lack of) political standing. Recommending more deliberation within the same kind of process appears problematic, given that it currently excludes many voices.

The 'problem' of 'distributed control' is also to be tackled by enrolling others to TM's aims – by 'stimulating local actors to reconsider basic assumptions about problems and solutions and their "normal" way of dealing with issues'. Conflict is to be kept 'within bounds' through 'modification of the self-understanding of actors (identities), strategic capacities and interests of individuals and collective actors and hence their preferred strategies and tactics' (Kemp *et al.* 2007, p. 320). Once the public, local actors and dissenting stakeholders have been enrolled in this way, 'reflexive' use of (centralising) processes such as 'functional coordination, cross-linking and integration' are noted as 'common strategies for dealing with problems of distributed control' (Kemp *et al.* 2007, p. 318).

Two strategies address the possibility that incoming governments might not continue TM efforts. Firstly, 'policy makers and politicians have to accept that a transition is needed'; and secondly, transition institutions should be adaptive and proceed gradually and reflexively in order to co-evolve with changing politics (Kemp *et al.* 2007, p. 318). TM remains unclear on political strategies for determining what to do in the short term to effect long-term change, in part because 'there exists little theory on how to do this', but a 'dual strategy of foreseeing and backcasting based on integrated system analysis may be useful here' (Kemp *et al.* 2007, p. 318). Any danger of 'lock-in' is avoided through two strategies: support 'should be given not just to one option but to a [dynamic] portfolio'; and a 'second strategy is simply to be prudent ... wait for better solutions and spend money on their creation' (Kemp *et al.* 2007, p. 319).

The issues above challenge the status of vanguard transition arenas in relation to other parts of government, parliament and wider policy institutions. How should the deliberations, activities and advocacy of the arena be advanced; indeed, is this desirable from a democratic point of view? Sources of legitimacy for transition arenas are not elaborated in the literature (Shove and Walker 2007, Hendriks 2008), but the innovativeness, vision and change agent qualities of each constituent member appear to be important. Whilst an ability to get things done can confer some legitimacy, however, this is rarely a sufficient source of political legitimacy. Nor does it address accountability issues. Related to this are questions about political and transition leadership. Strategic thinkers who bring their groups to transition arenas appear likely contenders, but what about political leaders who champion arena visions, or the bottom-up niche pioneers who lead the way in terms of sustainable behaviours? The TM approach does not really consider how to build a wider political legitimacy for authoritative leadership that can overcome political and economic resistance to its visions and niches.

Strategies to improve the political context for successful TM

Changing the political context within which TM operates need not be something done within TM. A government considering a TM approach might invest more public funds in relevant academic research and appraisal processes, and take other steps to elevate the standing of interactive, expert-led decision-making. Making wider general use of high-level expert assessments to justify policy (as with the UK's Stern Review on climate change) could also bolster the perceived legitimacy of TM initiatives. Another political tactic might be to strengthen commitments and links with political arenas that work in a similar way, for example the International Energy Agency, European Commission and European Environment Agency. Above all, however, the context must permit TM to drive horizontal and vertical integration with other parts of government and their policy networks. This is not easy, as the systems thinking at the heart of TM does not sit easily with existing policy institutions and departmental frameworks (Geels *et al.* 2008).

Many of the difficulties in implementing TM stem from an inadequate conceptualisation of power and agency (Smith *et al.* 2005). Having spotted niches as important sites for innovation and seeds for change, TM appears to emphasise niche-based agency to the exclusion of other processes, such as mass mobilisation, attempts to achieve discursive hegemony, and strategies based on political economy. Perhaps the lessons that TM has drawn from historic transitions, few of which have been politically led, should be augmented with lessons from the history of large, purposive political programmes.

Political strategies for transitions

In the reflexive spirit TM calls for, it is worthwhile questioning the assumption in TM (and this volume) that analysts should guide governments towards policies that avoid political fallout. Deciding between options remains, after all, a political calculation. Moreover, insights from the socio-technical transitions literature could equally be directed at entrepreneurs, consumers, communities, pressure groups and/or investors interested in low carbon transitions – governments will make few emissions cuts themselves: it is how they seek cuts by others that matters. Indeed, 'government' needs to be unpacked. One needs to consider, for example, whether a political strategy for transitions is to be developed by a political party while in office or opposition.

Winning office on a platform that included low carbon transitions as a central political project would lend significant legitimacy to subsequent efforts.

Approaching low carbon transitions as a political project suggests familiar strategies and tactics, such as creating large, powerful and well-funded institutions with a remit to pursue the project's aims. Other institutions' power might have to be curtailed, for example the power of government departments that have a close client relationship with powerful regime incumbents such as fossil energy companies. Steps could be taken to tie future governments into continuing the political project (Pierson 2000). The Climate Change Act in the UK, for example, commits UK governments to legally binding cuts in greenhouse gas emissions over the period up to 2050. This all implies a certain drive and readiness for conflict that bears little relation to TM's implicit model of politics.

The electricity regime in typical affluent democracies since the 1980s has had regulated competition as its main driver and organising principle. This is now perceived as problematic, and alternative agendas are being seriously considered (Scrase and MacKerron 2009). If the market model is rejected, governments face two options. They can either take a top-down policy approach that forces a transition to a low carbon society, or they can facilitate bottom-up momentum for change by empowering people to make their homes, communities and lifestyles sustainable. The former might take the form of a corporatist strategy in which governments accept that energy services will be supplied by a small number of large firms and try to enrol these firms to support and implement low carbon policies. Under such arrangements, however, governments would be under pressure to defend the interests of large energy companies, which implies that low carbon transition pathways are more likely to proceed by subsidising nuclear power and CCS than by supporting renewables.

In contrast, the alternative pathway, which would make much greater use of distributed and micro-generation, implies breaking up the large energy companies and reducing dependence on the national grid for electricity supplies. This route would presuppose a groundswell of popular concern about climate change and a readiness to use new technologies to cut emissions, combined with policy frameworks that enable this rather than making local pioneers continually face impossible odds. The corporatist strategy would derive its power base from industry and experts, while the decentralising strategy would be based on popular engagement and democratic support. Despite TM's emphasis on 'niches', in terms of political strategy it appears more closely aligned with corporatism than with radical decentralisation.

The decentralisation pathway might make use of transitions analysis, but quite differently from the ways sketched above. Transition analysis would be directed at making it as easy as possible for individuals, families and communities to invest, organise, link into low carbon networks of one kind and another, and so on. It is difficult to square that with policy generated in technocratic arenas through appraisal and foresight exercises. Moreover, it implies high levels of political commitment to pressure energy regimes accordingly. This kind of political project, underpinned by choices between contending green pathways, lies beyond TM.

Conclusion

One can argue that TM is a procedural tool that can be put to use by many different players. Yet no tool is neutral, and we have to consider whether the nature of TM renders it susceptible to capture. Does emphasis on consensus amongst an elite vanguard, a niche-based momentum for change, and reliance on integration with more powerful policy domains, really challenge the structures that TM hopes to transform? Even though TM proclaims participatory and reflexive processes, the narrow power base of its transition arenas, coupled with a limited and largely implicit political strategy, forces it towards technocratic strategies.

In principle, the open nature of TM and flexibility in purpose means that it might be possible to use it in ways that help empower people and facilitate a groundswell of bottom-up sustainability initiative (Seyfang and Smith 2007). There is certainly much to commend a multi-level, socio-technical analysis of how our needs are realised and how sustainable pathways might be realised in more democratic ways. But this would require a concomitant redistribution of resources to support the numerous, distributed and context-sensitive niches that would explore those visions and pathways.

TM has been a remarkable success in casting existing policy measures in an informative new light. However, in the context of the typical affluent democracy it is difficult to avoid the conclusion that the political strategies and tactics it advocates are inadequate for the task it has set itself. Yet the history of environmentalism reminds us that groups in society are perpetually trying to develop niche alternatives and pressure incumbent regimes in many different ways and with differing levels of agency and influence. A messy, informal transition politics already exists. In our view, this suggests possibilities for mobilisation in a political programme for low carbon transitions.

References

Berkhout, F., Smith, A., and Stirling, A., 2004. Socio-technical regimes and transition contexts. *In*: B. Elzen, F.W. Geels, and K. Green, eds. 2004. *System innovation and the transition to sustainability*. Cheltenham: Edward Elgar, 48–76.

Bush, G., 2007. *Text of a speech by the US President to the major economies meeting on energy security and climate change* [online]. US Department of State, 28 September 2007, Available from: http://www.whitehouse.gov/news/releases/2007/09/20070928-2.html

Dosi, G., 1982. Technological paradigms and technological trajectories. *Research Policy*, 11, 147–162.

Elzen, B., Geels, F.W., and Green, K., eds. 2004. *System innovation and the transition to sustainability*. Cheltenham: Edward Elgar.

Geels, F.W., 2004. From sectoral systems of innovation to socio-technical systems: insights about dynamics and change from sociology and institutional theory. *Research Policy*, 33, 897–920.

Geels, F.W., Monaghan, A., Eames, M., and Steward, F., 2008. *The feasibility of systems thinking in sustainable consumption and production policy: a report to the Department for Environment, Food and Rural Affairs*. Brunel University. London: Defra.

Guardian, 2008. Britain tries to block green energy laws. 24 July, 1.

Helm, D., 2005. The assessment: the new energy paradigm. *Oxford Review of Economic Policy*, 21 (1), 1–18.

Hendriks, C., 2008. On inclusion and network governance: the democratic disconnect of Dutch energy transitions. *Public Administration*, 86 (4), 1009–1031.

HM Government, 2008. *Building a low carbon economy: unlocking innovation and skills*. London: Department for Environment, Food and Rural Affairs.

Hoogma, R., *et al.*, 2002. *Experimenting for sustainable transport: the approach of strategic niche management*. London: Spon Press.

Hughes, T.P., 1983. *Networks of power: electrification in Western society, 1880–1930*. Baltimore, MD: Johns Hopkins University Press.

Kemp, R. and Loorbach, D., 2003. Governance for sustainable development through Transition Management. *In*: *Paper presented to the Berlin Conference on the Human Dimensions of Global Environmental Change*, 5–6 December.

Kemp, R., Schot, J., and Hoogma, R., 1998. Regime shifts to sustainability through processes of niche formation: the approach of Strategic Niche Management. *Technology Analysis and Strategic Management*, 10 (2), 175–195.

Kemp, R., Rotmans, J., and Loorbach, D., 2007. Assessing the Dutch energy transition policy: how does it deal with dilemmas of managing transitions? *Journal of Environmental Policy and Planning*, 9 (3), 315–331.

Loorbach, D., 2007. *Transition management: new mode of governance for sustainable development*. Utrecht: International Books.

Markard, J. and Truffer, B., 2006. Innovation processes in large technical systems: market liberalization as a driver for radical change? *Research Policy*, 35 (5), 609–625.

Meadowcroft, J., 2005. Environmental political economy, technological transitions and the state. *New Political Economy*, 10 (4), 479–498.

Meadowcroft, J., 2007. Steering or muddling through? Transition management and the politics of socio-technical transformation. *In*: *Paper presented to the Workshop on Politics and Governance in Sustainable Socio-technical Transitions*, 19–21 September, Schloss Blankensee.

Mitchell, C., 2007. *The political economy of sustainable energy*. Basingstoke: Palgrave Macmillan.

OECD (Organisation for Economic Cooperation and Development), 1999. *Managing national innovation systems*. Paris: OECD.

Persson, G. and Blair, T., 2003. *Open letter* [online]. Available from: http://www.defra.gov.uk/environment/business/envtech/pdf/blair-persson.pdf [Accessed 9 December 2005].

Pierson, P., 2000. Increasing returns, path dependence and the study of politics. *American Political Science Review*, 94 (2), 251–267.

Porter, T.M., 1995. *Trust in numbers. The pursuit of objectivity in science and public life*. Princeton, NJ: Princeton University press.

Quist, J., 2007. *Backcasting for a sustainable future: the impact after 10 years*. Delft: Eburon.

Raman, S. and Shove, E., 2000. The business of building regulation. *In*: S. Fineman, ed. *The business of greening*. London: Routledge, 134–149.

Rip, A. and Kemp, R., 1998. Technological change. *In*: S. Rayner and E.L. Malone, eds. *Human choices and climate change volume 2 – resources and technology*. Columbus, OH: Battelle, 327–399.

Rohracher, H., 2001. Managing the technological transition to sustainable construction of buildings: a sociotechnical perspective. *Technology Analysis and Strategic Management*, 13 (1), 137–150.

Rotmans, J., Kemp, R., and van Asselt, M., 2001. More evolution than revolution: transition management in public policy. *Foresight*, 3 (1), 15–31.

Scrase, J.I. and MacKerron, G., 2009. *Energy for the future: a new agenda*. Basingstoke: Palgrave Macmillan.

Seyfang, G. and Smith, A., 2007. Grassroots innovations for sustainable development: towards a new research and policy agenda. *Environmental Politics*, 16 (4), 584–603.

Shove, E., 1998. Gaps, barriers and conceptual chasms: theories of technology transfer and energy in buildings. *Energy Policy*, 26 (15), 1105–1112.

Shove, E. and Walker, G., 2007. CAUTION! Transitions ahead: politics, practice and sustainable transition management. *Environment and Planning A*, 39, 763–770.

Smith, A., 2007. Translating sustainabilities between green niches and socio-technical regimes. *Technology Analysis and Strategic Management*, 19 (4), 427–450.

Smith, A. and Kern, F., 2009. The transitions storyline in Dutch environmental policy. *Environmental Politics*, 18 (1), 78–98.

Smith, A. and Stirling, A., 2007. Moving outside or inside? Objectification and reflexivity in the governance of socio-technical systems. *Journal of Environmental Policy and Planning*, 9 (3–4), 351–373.

Smith, A., Stirling, A., and Berkhout, F., 2005. The governance of sustainable sociotechnical transitions. *Research Policy*, 34, 1491–1510.

Stirling, A., 2008. Opening up and closing down: power, participation and pluralism in the social appraisal of technology. *Science, Technology and Human Values*, 33 (2), 262–294.

Unruh, G.C., 2002. Understanding carbon lock-in. *Energy Policy*, 28, 817–830.

Unruh, G.C., 2005. Escaping carbon lock-in. *Energy Policy*, 30, 317–325.

Van der Brugge, R. and van Raak, R., 2007. Facing the adaptive management challenge: insights from transition management. *Ecology and Society* [online], 12 (2). Available from: http://www.ecologyandsociety.org/viewissue.php?id=68#Research [Accessed 11 August 2009].

Walker, W., 2000. Entrapment in large technology systems: institutional commitment and power relations. *Research Policy*, 29 (7–8), 833–846.

Weber, M. and Hemmeslkamp, J., eds. 2005. *Towards environmental innovation systems*. Berlin: Springer.

Networks, resources, political strategy and climate policy

Hugh Compston

This analysis is designed to show how policy network theory can be used to gain insights into the politics of climate change and climate policy. A version of policy network theory is set out based on the idea that policy networks are created and sustained by interdependencies among political actors. This theory identifies the main types of resources that are exchanged, and the main kinds of political actors that are likely to engage in resource exchange in the field of climate policy. Policy network theory is then used to unpack the main strategic options that are available to governments. The analysis concludes by listing 10 specific implications for governments that want to take more effective action against climate change while avoiding significant political damage.

Introduction

Governments are agreed that climate change is under way (IPCC 2007), we know many economically viable ways to reduce greenhouse gas emissions (Stern 2007), yet emissions are still rising. This suggests that the principal obstacles to quicker progress may be political, and that new political strategies need to be identified to enable activist governments to take more effective action against climate change while avoiding significant loss of political support.

My aim here is to apply a particular version of one of the dominant theories of policy making today, policy network theory, to the task of identifying political strategies that may make it easier for governments to take stronger action against climate change while avoiding significant political damage. The analysis begins by describing this version of policy network theory, which is based on the idea that policy networks are created and sustained by interdependencies among political actors. This discussion is followed by descriptions of the main types of resources that, according to this theory, are exchanged, and the main types of political actors that are likely to engage in resource exchange in the field of climate policy. In the next section political strategy is defined and policy network theory is used to unpack the main strategic options for governments that want to take stronger action on climate change. The final section draws out the ten main implications of this for identifying political strategies for governments that wish to take more effective action against climate change.

Policy network theory and resource exchange

The term 'policy network' has been used in political science at least since the 1970s, although the phenomena to which it refers had of course been described in other terms before then. In its most basic sense it refers to the set of political actors inside and outside of government who are involved in, or take an interest in, the making of public policy, and/or the relations between these actors. However, political scientists have also used the term in more specific ways, and a literature has grown up in which it is one of the central concepts, if not *the* most central concept. This literature, as one might expect from academic discourse, is rather disparate, so that there are many different versions of what exactly policy networks are.

One feature that many of these uses of the term share, however, is the idea that the relationships between network members are based on resource interdependencies: each actor wants something from one or more other actors and is prepared to exchange something of their own in order to get it. In her extensive survey of policy network literature, Tanja Börzel (1998, p. 254) concludes that although the network concept varies considerably between and within different disciplines:

> They all share a common understanding, a minimal or lowest common denominator definition of a policy network, as a set of relatively stable relationships which are of non-hierarchical and interdependent nature linking a variety of actors, who share common interests with regard to a policy and who exchange resources to pursue these shared interests acknowledging that co-operation is the best way to achieve common goals.

Perhaps the most prominent version of policy network theory in Britain is that put forward by Rod Rhodes. This picks up earlier American work in defining a policy network as 'a complex of organisations connected to each other by resource dependencies' (Benson 1982, p. 148) and views policy making as consisting largely of a process of exchanges of resources using specific political strategies within understood 'rules of the game' (Rhodes 1985, pp. 4–5). Similarly, Van Waarden (1992, p. 31) sees policy networks as arising from the interdependence of various actors: administrators need political support, legitimacy, information, coalition partners against bureaucratic rivals, and assistance with implementation, while interest groups want access to policy making and implementation, and concessions to their interests.

This leads to resource exchange that, over time, may become institutionalised into networks.

Resource dependency is also central to the more recent literature on governance. Rhodes, for example, defines governance as self-organising, inter-organisational networks and again focuses on power dependence and the exchange of resources using strategies within known rules of the game (Rhodes 2000, pp. 60–61). Klijn and Koppenjan (2000, p. 139) argue that the network approach presents public policy as the result of the interaction between a multitude of organisational actors each of which depends on other organisa-tions for resources and therefore needs to exchange resources in order to survive and achieve its objectives. Stoker (1998, p. 22) also sees governance as involving power dependence: '(a) organisations committed to collective action are dependent on other organisations; (b) in order to achieve goals, organisations have to exchange resources and negotiate common purposes; (c) the outcome of exchange is determined not only by the resources of the participants but also by the rules of the game and the context of the exchange'.

My aim here is to unpack the logic of this resource dependency version of policy network theory as it relates to the politics of climate change and climate policy. Policy network theory is therefore defined and delimited as the ensemble of two elements. The first of these is a definition of a policy network as 'a set of political actors who engage in resource exchange over public policy (policy decisions) as a consequence of their resource interdependencies'. Note that this definition includes within it a causal dynamic, namely the dependence of political actors on each other for resources that pushes them to engage in resource exchange over public policy.

The second element consists of those propositions that are either presupposed or entailed by this definition, as follows.

There are policy decisions

As a theory about how policy decisions are made, policy network theory obviously presupposes the existence of policy decisions, understood as authoritative choices about policy instruments and their settings. This in turn presupposes the existence of legal rules that specify who is authorised to take decisions in any given area, under what conditions and using which procedures.

There are individuals and/or groups that possess (perceived) tradable resources

Defining policy making in terms of resource exchange presupposes that there are individuals or groups (policy actors for short) who possess resources that can be exchanged. To some extent the nature of tradable resources can be deduced from the nature of policy decisions. If there are policy decisions that are legally binding then there must be an actor or actors possessing the legal authority to take these decisions and therefore the capacity to exchange policy amendments for resources that this actor or actors want, such as the formal approval of other public actors, cooperation with implementation, and political support.

Policy actors have distinct policy preferences

Engagement in resource exchange presupposes that network members have distinct preferences about what the content of policy decisions (policy instruments and settings) should be. Otherwise why bother engaging?

There are (perceived) policy problems and solutions

Policy preferences and the existence of resource exchange presuppose that one or more actors consider that there are problems in the world to which policy decisions can provide solutions.

However, the complexity of reality and the limited cognitive resources of human beings mean that understanding the causal relations in the world that underlie conceptions of problems and solutions must necessarily involve simplifying and relating new information to existing ideas in order to produce meaningful and structured interpretations. Sociological institutionalists argue that information is selected and processed by cognitive structures (variously named schemas, frames or inferential sets) that determine 'what information will receive attention; how it will be encoded; how it will be retained, retrieved and organised into memory; and how it will be interpreted, thus affecting evaluations, judgments, predictions and inferences' (Scott 2001, p. 38).

What this means is that the problems as identified by policy actors are not objective descriptions of events and conditions in the world but rather perceptions of these, defining 'perceptions' in the dictionary sense of 'ways of regarding, understanding, or interpreting' (*Concise Oxford Dictionary*). Proposed solutions to these problems, of course, are constructions from the start because they do not purport to describe situations in the world but rather refer to what an actor claims would happen if a certain policy decision was taken.

Policy actors have strategies designed to maximise their chances of realising their policy preferences

If resource exchange is to be used by a policy actor to help them achieve their policy preferences, it follows that they must have a strategy, defined as a plan of action designed to maximise their chances of realising these preferences.

There are incentives for policy actors to regulate their interaction

To the extent that policy actors remain dependent on each other for resources, they have an incentive to exchange resources on a continuing basis and therefore an incentive to establish mutually recognised procedures (legal rules or informal norms) that facilitate interaction.

To sum up, policy network theory as defined holds that policy change is largely determined by resource exchange involving political actors and their resources, preferences and strategies, and that these in turn are necessarily influenced by perceptions of problems and solutions and, more contingently, by policy network-specific rules and norms.

Resources and resource exchange

As resource exchange is central to the resource dependency version of policy network theory, it is vital to be quite clear about what this means before going on to look at strategy.

Theories of power dependency and resource exchange have a longer history in sociology and social psychology than in political science (see, for example, Emerson 1962, Blau 1964). Although analyses in these traditions pay little attention to policy networks as such, they do make some relevant points.

In particular, power is conceptualised in terms of dependency. The argument is as follows.

> A depends on B if he aspires to goals or gratifications whose achievement is facilitated by appropriate actions on B's part. (Emerson 1962, p. 32)

Where this is the case, B is in a position to grant, deny, facilitate or hinder A's gratifications, putting B in a position to make demands of A that A has to accede to if he or she is to obtain gratification.

> Thus, it would appear that the power to control or influence the other resides in control over the things he values ... In short, *power resides implicitly in the other's dependency*. (Emerson 1962, p. 32)

Emerson then distinguishes between two dimensions of dependence:

> The dependence of actor A upon actor B is (1) directly proportional to A's *motivational investment* in goals mediated by B, and (2) inversely proportional to the *availability* of those goals to A outside of the A–B relation. (Emerson 1962, p. 32)

Resource dependency, then, means that a policy actor wants or needs something that is controlled by another actor. Resource exchange presupposes that the relevant resource is transferable. Consequently, we can define a resource in the context of policy network theory as being anything that (a) is controlled by a policy actor, (b) is desired by another policy actor, and (c) can be transferred or exchanged in some relevant sense.

The political science and corporate management literatures are replete with references to political resources in the sense of characteristics that help political actors get their own way (see, for example, Dahan 2005, Scharpf 1997, p. 43, Pappi and Henning 1998, p. 557, Rhodes 2006, p. 431). But not all of these are tradable in any direct sense. Public actors, for example, do not formally transfer legal authority to the actors with which they are exchanging resources. What are really being exchanged for other resources are policy amendments. The significance of legal authority is that it is what enables the relevant public actor to trade policy amendments: it is an enabling resource rather than a tradable resource.

For this reason it is necessary to distinguish tradable resources from other types of resources. Table 1 sets out a list of political resources based on surveying the relevant literature with a view to identifying exactly what it is that is being exchanged, and grouping the resources identified into a manageable number of general categories (Compston 2009, ch. 2).

It is important to note that the resources listed are arranged according to the type of actor that controls them *before* any resource exchange takes place. *After* resource exchange the situation may be quite different: private actors, for example, may have received the policy amendments that they want while public actors may have received the resources that they want, such as political support.

It is also important to note that the exact terms of any resource exchange depend not only on the resources that each party to the exchange actually controls but also on how they are perceived: threats by firms to disinvest, for example, are only political resources if they are believed.

Furthermore, by their nature each of these types of resources has a reward side and a punishment side. Public actors may reward other actors for their cooperation by granting them policy concessions, for example, or may punish them by withholding these or even by making the policy less palatable to these actors than it was originally. By the same token private actors may provide public actors with political support if they get the policy

Table 1.   Main tradable resources of policy network members.

Controlled by:	Resource	Comments
Public actors alone	Policy amendments	The perceived need to grant policy amendments depends on the perceived need for other resources
	Access	Being granted access creates the opportunity to persuade
Public and private actors	Veto power	
	Information	
	Cooperation with implementation	
	Recourse to the courts	
	Political support	
	Patronage	
Private actors alone	Private investment	
	Fluid funds	

amendments that they want, or may mobilise against the government if they do not.

Let us have a closer look at these tradable resources, starting with those that only public actors possess.

Policy amendments

By 'policy amendment' as a tradable resource is meant a change in the choice of policy instrument(s) and/or settings in areas such as regulation, funding and taxation in the direction desired by the actor with whom the relevant resource exchange is taking place. Only public actors with the legal authority to make binding decisions in the relevant policy area are in a position to use policy amendments as a tradable resource, which means that legal authority is the enabling resource.

One complication here is that the legal authority to make decisions in a certain policy area may be divided among two or more public actors. In such cases policy amendments may be traded among these actors in order to reach agreement on the decision.

Access

By 'access to policy making' is meant inclusion on distribution lists, two-way communication with officials and/members of the political executive, inclusion on relevant committees, and in general the opportunity to submit arguments in the knowledge that they will be considered. Access is valuable for policy actors because it gives them information about what public actors are doing and thinking, plus the chance to put their arguments to these public actors. In some cases access is mandated by law. Otherwise the power to grant or deny access is the prerogative of the public actors that possess the legal power of decision in any given policy area.

The next few types of tradable resources may be held and exchanged by both public and private actors.

Veto power

Policy actors with veto power, such as opposition parties that control powerful upper houses of legislatures, may use this to block decisions unless certain policy amendments are made. This means that the precise tradable resource consists of refraining from exercising their veto power.

Information

There are at least three ways in which information can be used as a resource. Actors with key information that they and only they control may exchange this directly for policy amendments. Information may influence policy decisions directly by changing the policy preferences of the public actor(s) with the legal authority to make the relevant policy decisions when these decisions are taken without engaging in resource exchange. Policy learning may also alter the nature of any resource exchange that does take place, and thus the policy decision that emerges, via altering the policy preferences of one or more actors.

Cooperation with implementation

There is no point in adopting a policy if it cannot be implemented properly, so where actors have the capability to impede or block implementation, public actors have an incentive to exchange policy amendments for cooperation with implementation.

Recourse to the courts

Where public or private actors are in a position to go to court to block or at least delay a policy decision or its implementation, refraining from using this legal option becomes a resource that can be traded for policy amendments.

Political support

The type of policy actor most in need of political support is the political executive, that is, the leadership of the party or parties in government, as governments in democratic countries must submit themselves to elections every few years.

Political support as a tradable resource comes in a number of forms. First, and following Pappi and Henning (1998, p. 557), private actors may be in a position to mobilise the public or specific groups in favour of policy proposals, or at least to refrain from mobilising them against policy proposals. Political mobilisation in this sense includes media campaigns, strikes, petitions and demonstrations.

Second, public actors may seek the approval of relevant private actors even where political mobilisation is not an issue in order to give the chosen policy more (apparent) legitimacy. Here the enabling resource for such private actors is a good reputation in the eyes of the groups and voters that public actors wish to get or keep onside.

Third, members of the political executive are individually and collectively dependent on the support of other politicians to stay in office. The government as a whole needs the support of the legislature. Ministers need the support of the head of government. The head of government needs the support of the party. As a consequence legislators outside the government may be able to trade political support for policy amendments, the head of government may trade his or her support for policy amendments by ministers, and the party as a whole may trade political support for policy amendments by the political executive.

Fourth, where the public is aware of an issue and cares, public actors may in effect deal directly with voters by making policy amendments designed to elicit higher opinion poll ratings in return.

The significance of political support as a tradable resource depends on how much the relevant public actor wants or needs it. A minister who is popular with colleagues and sits in a government with a secure legislative majority that is riding high in the opinion polls, for example, does not need to trade policy amendments for more political support as much as an unpopular minister in a minority government facing an election well behind in the polls.

The significance of political support as a tradable resource also depends on perceptions: whether environmentalists can mobilise voters against a particular government, for example, is uncertain ahead of its being demonstrated (or not), so its value as a tradable resource prior to this depends on whether the relevant public actors believe that environmental organisations have this capacity.

Patronage

Both public and private actors may exchange patronage for other resources. Public actors may grant positions in the government or administration, or honours, in exchange for benefits such as investment, campaign contributions or bribes. Private actors, especially corporations, may give ex-ministers, ex-officials and/or their family members positions in their own or related organisations, such as well-paid directorships, in exchange for policy amendments while they are still in office. Both forms of patronage constitute corruption: the exchange of personal favours for policy amendments or other benefits.

Private investment

As Lindblom points out (1977, ch. 13), governments are held responsible for economic success but do not control the investment needed to ensure this. For this reason decisions about the location and type of private investment can be used by firms as a tradable resource. Perceptions are very important here: a threat to disinvest, for example, is only effective if the relevant public actors believe that it would be carried out if they refuse to give the relevant firm(s) what they want.

Fluid funds

Fluid funds – cash and other easily transferred financial assets – can be used as a tradable resource in a number of ways. They can be used to bribe politicians and officials to grant policy amendments. They can be traded in the form of campaign contributions for specific policy commitments. And they can be used as an enabling resource to buy expertise, lobbying services and other tradable and enabling resources.

Network members

The conscious agents within policy networks are individual men and women, most of whom are acting as representatives of organisations of some sort. In abstract discussions of policy networks it is convenient to identify these organisations as the policy actors, but in reality what we have are 'composite actors' in which intentional action above the level of the individual is produced by internal interactions (Scharpf 1997, p. 52). We might add the observation that the nature of composite actors changes over time, so that what we have in fact are evolving composite actors.

The wide ramifications of climate change mean that a wide range of policy actors take an interest in the problem and possible policy responses to it. Among the most prominent of these are environment and economic ministers together with related departments and agencies, big energy producers and users, and climate change scientists. It is also worth noting that much European decision making on climate policy now takes place not at national level but at EU level and, to a lesser extent, at global level with agreements such as the Kyoto Protocol. Among other things this implies that national climate policies are influenced not only by domestic actors but also by the preferences, resources and strategies of foreign governments, and thus indirectly by whatever actors influence these governments. Finally, the fact that democratic governments (political executives) live under the shadow of the next election brings the electorate into play as a sort of quasi-policy actor in that public actors may in effect deal directly with voters by making policy amendments designed to elicit higher opinion poll ratings in return.

Table 2 lists the main types of policy actors that would be expected to be members of national-level policy networks relating to climate change.

Policy networks, political strategy and climate policy

If resource exchange is to be used by policy actors to help them achieve their policy preferences, it follows that they must have a strategy, defined as a plan of action designed to maximise their chances of realising their policy preferences. Since policy network theory specifies that policy decisions are determined mainly by resource exchange, it follows that actors' strategies must include decisions about how available resources are to be deployed. Strategies must also include communicating to other actors their policy preferences, diagnoses of problems and solutions, and (at least to some extent) their intentions in relation to deployment of resources, as well as the collection of information about others actors' preferences, resources, conceptions of problems and solutions, and strategies. Forms that this communication may take include provision of information, argument, bargaining, and the making of threats and promises.

Table 2.   Generic policy network relating to climate change.

Type of actor	Specific actor
Political executives and associated parties	Chief executives Environment ministers Economic, finance, industry, energy, transport and technology ministers
Non-government politicians	Opposition economic, environment, energy, transport, trade and technology spokespeople Legislative committees on the economy, environment, energy, industry, transport and technology Regional and local governments European Commissioners in areas relating to the economy, environment, industry, energy, transport and technology Foreign governments
Civil servants, special advisers and other public sector employees	Environment departments and agencies Economic departments and agencies, in particular energy, finance, industry, transport and technology European Commission Directorates-General relating to the economy, environment, industry, energy, transport and technology Regional and local administrations International organisations
Judges and regulators	Environmental, energy and transport regulators
Business	Fossil fuel producers and electricity generators: oil, gas, coal Renewable energy producers and electricity generators Nuclear energy producers and electricity generators Industrial users of energy such as steelworks Transport manufacturers and providers: road, air, rail, water Energy efficiency producers and services Trading firms of all sorts
Interest groups and non-government organisations	Business groups relating to energy, transport and industry Environmentalists Trade unions Motoring organisations
Media	National press and television; specialist environment, energy, transport and industry media
Experts	Climate change scientists, economists, and environmental, energy and transport experts in universities, think tanks and elsewhere
Electorate	Energy users, travellers, consumers of products produced using energy

In terms of resource exchange and other policy network variables, at least five main types of political strategy can be identified:

- Unilateral decision making without resource exchange.
- Simple resource exchange within existing network parameters.

- Moves to facilitate agreement on policy by facilitating resource exchange.
- Moves to change the policy preferences of other actors.
- Moves to change the terms of resource exchange.

This section draws out at least some of the implications of each of these strategic options for governments looking for ways to strengthen climate policies while avoiding significant political damage.

Unilateral decision making

Unilateral action is what many of those demanding more vigorous action on climate change tend to expect: the government works out what is technically and economically viable and then goes ahead and makes it happen.

But there is a problem. Most of these policies impose costs on someone and at least some of those affected are likely to try to block or weaken whatever policies impose these costs and to work to remove governments that introduce them. Such opponents are likely to include those industries that currently produce the lion's share of industrial greenhouse gas emissions and which therefore would be the most affected by the strengthening of climate policies. It is also clear that voters object to policies that obviously impose additional costs on them, such as higher fuel taxes and road pricing.

In terms of resource exchange what this means is that governments that take unilateral action are likely to lose any political resources that they possess by virtue of previous resource exchanges with these actors.

If we examine the resources that such actors could withhold from the government, however, it appears that the only resources that, in general, governments really could not do without are political support, cooperation with implementation, and private investment. Legal vetoes are a problem for government in countries such as the US, where there is a constitutional separation of powers, but much less of a problem in others, such as Britain. Policy actors might withhold information from the government, but it is not clear that this would create significant problems. Court action will only succeed if governments have exceeded their authority, which is unlikely to apply to many climate policies. And patronage and cash could be withheld, but this would only impose serious costs on governments that are quite corrupt. It is to be hoped that this is not relevant in very many cases.

For democratic governments, however, loss of political support could lead to the prime movers of more radical climate policies losing office as a result of a successful leadership challenge, defeat in a legislative vote of confidence, or defeat at the next election. Worse still, this could come at the hands of opponents who have pledged to reverse any new climate policies in order to win, so that any noble sacrifice would be in vain. Withdrawal of cooperation with implementation would wreck policies such as those designed to expand energy production from renewable sources. Loss of private investment, in the form of short-term capital flight or more long-term drying up of new real investment in production facilities and/or transfer of existing investment to other countries, would cause economic problems that in turn would be likely to damage governments' prospects of re-election.

Conversely, of course, providing that a proposed climate policy does not threaten significant loss of political support, cooperation with implementation or business confidence, unilateral action would be a viable option.

Tactics that could be used by governments that do decide to take unilateral action include:

- introducing unpopular policies during the early years of an administration to allow time for opposition to subside and for the benefits of such policies to become apparent before the next election arrives;
- targeting measures on particular industries while leaving others alone in order to make progress while keeping most of business onside, although the extent to which governments can do this may be limited by norms of equity of treatment, and if the government targets your industry today, it might be mine tomorrow;
- targeting economic sectors that are able to pass on at least a proportion of their extra costs to consumers, as this may facilitate the internalisation of environmental costs without government being blamed directly, although the media is often quick to publicise how carbon/energy taxes, for example, lead to higher prices for consumers; and
- adopting policies that target losses on small sections of society, and in particular on those groups that are least able to inflict political damage via the ballot box, impede implementation, or withdraw investment from the country.

Simple resource exchange within existing network parameters

If the main problem with unilateral action on climate policy is loss of political support, cooperation with implementation and business investment, the obvious alternative is to trade policy amendments for these resources, but only the minimum of policy amendments.

In a situation in which public policy is being determined by resource exchange among interdependent actors with distinctive policy preferences, strategy can be conceptualised, as mentioned above, as consisting of (1) communication of preferences and associated argumentation, (2) the selection, combination and sequence of resource deployments, and (3) whatever other actions are necessary to make this happen.

What each actor does is, of course, influenced by what others do: players 'gear their actions and the objectives which they pursue to the strategic behaviour and objectives of other actors. A strategy is thus a cohesive series of actions whereby one's own desires and ambitions are linked to the assessment of the desires and ambitions of other actors' (Klijn *et al.* 1995, p. 440).

Within this general framework there are at least two major strategic choices that public actors bent on further action on climate change need to make.

First, public actors need to decide which actors must agree if significant losses of political support, cooperation with implementation, and investment are to be avoided, as in general we would expect that obtaining the agreement of more than the minimum number of actors would result in more significant policy concessions being made and thereby weaker climate policies.

Second, governments need to decide exactly what adjustments to their ideal policies they are prepared to make in return for the resources that they seek, bearing in mind that too many concessions would sabotage the policy concerned. This of course presupposes judgements about what policy amendments are necessary in order to keep business and the electorate sufficiently onside. If the policy concessions required for this are judged to be unacceptable, of course, the only option for an activist government would be to impose the relevant policy and accept any consequences in terms of political support, policy implementation and/or business investment.

Where policy amendments are made, these can relate either to the climate policy under consideration or, as part of a package, to policies elsewhere that affect the same political actors. Corporations, for instance, are affected not only by climate policies but also by policies in areas such as taxation, business regulation and labour law. Thus policy linkages may provide ways in which opponents of particular climate policies can, in effect, be bought off.

Moves to facilitate agreement on policy by facilitating resource exchange

As well as playing the game within the prevailing network parameters, public actors who wish to obtain the agreement of other actors on climate policies have the option of trying to manage the network with the aim of facilitating resource exchange by means such as selective activation of the actors necessary to tackle a particular task; arranging interaction (formalising the agreements and rules that regulate interaction, for example by establishing conflict regulating mechanisms); matching problems, solutions and actors; facilitating interaction by making sure that appropriate procedures are in place; and mediation and arbitration (Kickert and Koppenjan 1997, pp. 47–51). Of these the most relevant to a government trying to secure agreement on stronger climate policies appear to be moves to establish procedures that facilitate easy interaction with other actors by means such as creating committees, putting relevant individuals in touch, and establishing norms of interaction so that actors know what they can expect from each other, for example clear recognition of agreements if and when they are reached.

Moves to change the policy preferences of other actors

A fourth strategy is to try to change the policy preferences of other actors by altering their perceptions of problems, solutions or other aspects of the policy process. Here of course the role of the media is central, as almost all political communication aimed at mass audiences is received as coverage in the media rather than direct and unmediated. This means that the nature of media coverage is seen, rightly or wrongly, as being critical in shaping public attitudes to climate change and climate policy (see Gavin, this volume).

Perhaps the most obvious communication strategy is the provision of information about climate change and the threat it poses, along with information about effective and practical responses. Another tactic is to stress the contribution of proposed climate policies to the achievement of other social and economic objectives such as energy security and employment. Messages aimed at citizens need to be simple and clear, which implies focusing on just a few selected indicators of climate change and its impacts, along with a small number of proposed solutions, and deploying metaphors and analogies to make it easier for citizens to understand complex ideas. Messages also need to be tailored to particular audiences and repeated as often as possible (Pralle, this volume).

Related to these points is the idea that political actors compete to secure support for their definition of reality by telling stories in which narrative devices such as plot and characterisation are at least as important as evidence and logic (see, for example, Hajer 1995). Issues that need exploring here include how to devise frames for climate change and climate policies that are credible and salient, and how best to promote these frames. Recent innovations in this area include the concept of a Green New Deal, while in the US the Apollo Project likens the task of controlling climate change to the effort during the 1960s to get a man on the moon (Pralle, this volume).

In relation to changing policy preferences it is also worth mentioning that if climate change is really under way we can expect an increase in the incidence of weather-related natural disasters. To the extent that a disaster of this sort can be plausibly linked in the public mind with climate change we would expect public concern about climate change to increase and lead to increased public demand for appropriate climate policies. We have already seen this in relation to events such as Hurricane Katrina in the US. The strategic significance of this is that from time to time we can expect windows of opportunity to open during which governments can put through more vigorous climate policies at a lower cost in terms of political support than they would suffer at other times. For this reason any government which is serious about climate change will keep fully-prepared policy options to bring out each time that there is a spike in public concern about climate change.

Moves to change the terms of resource exchange

In his analysis of resource dependency, Emerson (1962, p. 35) identifies a number of options for actors who wish to change the balance of power between themselves and other actors. Applying this analysis to the case of public actors wanting to change the balance of power within a policy network defined in terms of resource interdependencies implies that such actors have four main options.

(1) Reduce motivational investment in the resources controlled by other actors

By this is meant caring less about whether these resources are forthcoming. One way of becoming less dependent on resources controlled by others would be to switch to policy goals that require fewer resources from other actors. Governments may be able to avoid legislative veto points, for example, by prioritising climate policies that do not require legislative approval. As this strategy means following the path of least resistance, however, it may well be that it is already being used by governments to its maximum extent.

(2) Cultivate alternative sources for obtaining these resources

This means seeking political support, help with implementation and business investment from new sources. New sections of the electorate might be cultivated to compensate for losses in electoral support due to the introduction of new climate policies, for example, while policies that would be crippled by lack of cooperation on implementation could be replaced by policies for which cooperation is forthcoming.

(3) Get other actors to increase their motivational investment in the resources that you control

One relevant move here would be to hold out the prospect of a favourable policy change elsewhere, such as a tax cut, if cooperation is forthcoming. Another would be to acquire new resources such as additional authoritative powers with which to coerce other actors. Moves by national governments to acquire additional legal powers over planning permission, for example, as happened recently in Britain, increase the resources of governments by enabling them to offer policy concessions to firms that a more restrictive or localised planning system would not allow, which might lead to agreement being secured in return for fewer other policy concessions than would otherwise be the case. Another possible move would be to take full control of selected private firms by nationalising them. Such moves are out of fashion at present, or were until the recent financial crisis, but extensions of regulatory powers may also have this effect. An alternative possibility would be to form state agencies to undertake needed tasks that private firms are simply not carrying out, such as building coal-fired power stations that are equipped with carbon capture and storage.

(4) Deny to other actors alternative sources for obtaining the resources that they want from you

This implies actions such as imposing stricter controls on international transactions in order to make it harder for firms to shift investment elsewhere if they object to certain climate policies, although this particular move seems unlikely at present in view of current international trade rules. Another possibility would be to nurture cross-party consensus on climate change in order to limit the extent to which business groups or voters who object to certain climate policies are able to shift their political support to parties that oppose these measures.

A fifth strategic option derives from the Dutch school of network analysis (Klijn *et al.* 1995, p. 448).

(5) Alter the structure of policy networks

There are a number of possibilities here. First, public actors may introduce new actors and/or exclude others. This I interpret as giving new actors access to the policy-making process – consultations, committees and so on – while excluding others, since our definition of policy networks means that network membership as such is determined by resource interdependence rather than formal access. In at least some cases granting access may create a more cooperative attitude on the part of the newly included actor which may increase the chances of their accepting new climate policies. While excluding an actor implies a more antagonistic relationship, which may have negative consequences if they can influence public support, implementation or business investment, in at least some circumstances excluding an actor may make it easier for public actors to introduce contested climate policies by removing any expectation that these must be agreed by the actor concerned.

Second, public actors may try to alter relations between actors by establishing or changing organisational arrangements such as consultation procedures and advisory bodies. In terms of our resource dependency version of policy network theory this means altering policy network-specific rules and norms. However, it is not clear that such moves would be likely significantly to affect the likelihood of critical actors agreeing to climate policies that they had previously rejected.

Third, public actors may move to alter the distribution of resources by means such as improving information systems, recognising an organisation as an interlocutor, giving it access to permanent consultative bodies, granting subsidies, or granting it a legal monopoly. As mentioned above in relation to introducing new actors to the network, this may elicit greater cooperation from the actor(s) receiving these benefits.

Fourth, public actors may change interaction rules, such as decision-making procedures, and thereby alter the distribution of resources by means such as altering which particular public bodies hold the formal power of decision. Moving responsibility for energy from an economic to an environmental ministry, for example, should shift the power of decision towards actors who would be expected to be more likely to favour strong action to reduce energy-related emissions than their colleagues in the economic ministry.

Lessons

The aim of this analysis has been to draw out some of the implications of a resource dependency version of policy network theory for identifying political strategies for governments that wish to take more vigorous action against climate change while avoiding serious political damage. The ten main implications that have emerged are as follows.

(1) Democratic governments that want more action on climate change cannot do without political support, cooperation with implementation and the maintenance of business investment. For this reason an all-out unilateral strategy is impractical. Conversely, however, unilateral action is a viable option where a proposed climate policy does not threaten significant loss of political support, cooperation with implementation or business confidence.

(2) Tactics that can be used by governments that decide to take unilateral action include introducing unpopular policies early in their terms of office, imposing measures on particular industries while leaving other industries alone, targeting sectors that can pass on any additional costs, and targeting those social groups that are least able to retaliate via the ballot box.

(3) The main strategic choices for public actors operating within established parameters for resource exchange involve first identifying the minimum number of actors whose agreement is deemed to be necessary if unacceptable political damage is to be avoided, and then deciding exactly what policy concessions they are prepared to make in order to secure that agreement.

(4) It may be possible to buy off opponents of particular climate policies by putting together package deals that include amendments to policies elsewhere.

(5) It may be possible to improve the chances of securing an acceptable agreement by fine-tuning the formal and informal procedures that structure interaction among policy actors. Altering the distribution of formal decision-making power within government, for example by moving responsibility for energy policy from an economic ministry to an environment ministry, may reduce the strength of opposition within government to taking more vigorous action against climate change. Giving new actors access to the policy-making process may make them more likely to accept proposed policy changes, while excluding actors may make action more likely by removing any presumption that their agreement is a prerequisite.

(6) The acquisition of new legal powers may enable governments to offer more to companies and other actors and thereby reduce the policy concessions necessary to obtain their agreement to new climate policies.

(7) Cross-party agreement on climate policies should enable more radical policies to be put through because in such circumstances voters who oppose these policies have nowhere to go.

(8) It may be possible to alter the policy preferences of other actors by means of communications strategies ranging from providing information and stressing how climate policies serve other objectives as well, such as energy security, to adoption of discourses that frame the issue of climate change in new ways.

(9) If climate change is really under way we can expect increasingly frequent and serious weather-related natural disasters that are likely to lead to increased public demand for appropriate climate policies that temporarily reduces the political costs that such policies would otherwise incur and thereby opens repeated windows of opportunity for governments that keep fully-prepared policy options ready to go.

(10) Opportunities for strategic action may be opened by seeking new sources of political support, help with implementation and business investment to replace any lost through imposing climate policies on those actors that currently provide these resources.

References

Benson, J.K., 1982. A framework for policy analysis. *In*: D.L. Rogers and D. Whetten, eds. *Interorganizational coordination: theory, research and implementation*. Ames: Iowa State University Press, 137–176.

Blau, P., 1964. *Exchange and power in social life*. London: John Wiley.

Börzel, T.A., 1998. Organizing Babylon – on the different conceptions of policy networks. *Public Administration*, 76, 233–273.

Compston, H., 2009. *Policy networks and policy change*. Basingstoke: Palgrave Macmillan.

Dahan, N., 2005. A contribution to the conceptualization of political resources utilized in corporate political action. *Journal of Public Affairs*, 5, 43–54.

Emerson, R.M., 1962. Power-dependence relations. *American Sociological Review*, 27 (1), 31–41.

Hajer, M., 1995. *The politics of environmental discourse*. Oxford: Oxford University Press.

IPCC (Intergovernmental Panel on Climate Change), 2007. *Climate change 2007*. Cambridge: Cambridge University Press.

Kickert, W.J.M. and Koppenjan, J.F.M., 1997. Public management and network management: an overview. *In*: W.J.M. Kickert, E.-H. Klijn, and J.F.M. Koppenjan, eds. *Managing complex networks: strategies for the public sector*. London: Sage, 35–61.

Klijn, E.-H. and Koppenjan, J., 2000. Public management and policy networks. *Public Management*, 2 (2), 135–158.

Klijn, E.-H., Koppenjan, J., and Termeer, K., 1995. Managing networks in the public sector: a theoretical study of management strategies in policy networks. *Public Administration*, 73, 437–454.

Lindblom, C., 1977. *Politics and markets*. New York: Basic Books.

Pappi, F.U. and Henning, C.H.C.A., 1998. Policy networks: more than a metaphor? *Journal of Theoretical Politics*, 10 (4), 553–575.

Rhodes, R.A.W., 1985. Power-dependence, policy communities and intergovernmental networks. *Public Administration Bulletin*, 49, 4–31.

Rhodes, R.A.W., 2000. Governance and public administration. *In*: J. Pierre, ed. *Debating governance*. Oxford: Oxford University Press, 54–90.

Rhodes, R.A.W., 2006. Policy network analysis. *In*: M. Moran, M. Rein, and R.E. Goodin, eds. *The Oxford handbook of public policy*. Oxford: Oxford University Press, 425–447.

Scharpf, F., 1997. *Games real actors play*. Oxford: Westview Press.

Scott, W.R., 2001. *Institutions and organisations*. 2nd ed. London: Sage.

Stern, N., 2007. *Stern review: the economics of climate change*. Cambridge: Cambridge University Press.

Stoker, G., 1998. Governance as theory: five propositions. *International Social Science Journal*, 50 (155), 17–28.

Van Waarden, F., 1992. Dimensions and types of policy networks. *European Journal of Political Research*, 21 (1), 29–52.

Political strategy and climate policy: a rational choice perspective

Frank Grundig

Rational choice models are employed to explain both the formation of states' domestically derived negotiation positions on climate change and the dynamics of these international negotiations. This analysis leads to the identification of a number of promising political strategies: taking steps to enlarge the membership of environmental non-governmental organisations; assessing whether the resources of these organisations would be more effectively spent on campaign contributions rather than other activities; using organisations such as the World Bank to assist developing countries to strengthen civil society in these countries; stepping up information campaigns; re-balancing abatement costs between the EU and the US; and making abatement more efficient by introducing an international emissions cap and trade scheme.

Introduction

The outcome of negotiations to prevent climate change is shaped by both the positions of national governments and the dynamics of the international negotiation process. Here we employ a number of rational choice perspectives on domestic politics to illuminate how national negotiating positions are formed. The rational choice perspective seeks to explain political outcomes based on a small number of assumptions and deductive reasoning. Rational choice models usually assume that actors are self-interested and that they maximise their utility in accordance with their preferences. We then plug the findings into a rational choice model of negotiations over public goods in order to identify a number of *domestic* political strategies that may allow more effective *international* agreements to be reached. The central argument developed below suggests that for a number of domestic politics reasons, such as the variation in net benefits from abatement or special interest group activity, states differ with regard to their desire to combat global climate change, making it difficult to reach effective international agreements. We argue that it is nevertheless possible for governments to achieve a more effective agreement by moving away from the current Kyoto Protocol provisions, and for civil society actors to effect some change if citizens can overcome collective action problems.

The analysis begins with a domestic political economy perspective, which assumes that states act as unitary utility maximising actors, in order to shed light on what determines states' policy positions. Drawing on economic models of climate change, this perspective shows that the Kyoto Protocol would have led to a distribution of net benefits that made US participation unlikely. We then disaggregate the state and ask whether and how special interests, industrial or environmental, may move a state's policy positions away from maximising national welfare. This approach suggests that policy positions might reflect a combination of the policy maker's own preferences, strategic voters' preferences, and special interest groups' preferences. We then discuss a rational choice model of negotiations which attempts to explain the (in)effectiveness of the climate change regime in terms of divergences among national positions and the ability of important states to make side-payments to achieve progress. Including the international dimension is particularly important in the case of climate change for at least three reasons. First, no one state can address the issue of climate change unilaterally. Second, a stable climate is a relatively pure public good, which invites free-riding on others' efforts to reduce emissions. Third, any unilateral domestic action might lead to carbon leakage since under current GATT/WTO rules a state cannot compensate for another state's comparative advantage by increasing tariffs (Conconi 2003).[1] Based on a combination of the negotiation model with the domestic approaches, we outline a number of policy recommendations, including a re-balancing of implementation costs between the US and the EU in any newly negotiated protocol, attempts to make the regime as economically efficient as possible, efforts to increase the influence of environmental non-governmental organisations (ENGOs) through membership drives, more vigorous information campaigns, and re-evaluations of the most effective way to spend 'green' money.

The political economy of climate change at the state level

We first assess the costs and benefits that states incur if we treat them as unitary, welfare maximising actors. In the next section we relax this assumption and look at policy makers' maximisation of welfare.

Perhaps the most useful basic characterisation of an 'interest based explanation of international environmental policy' was put forward by Sprinz and Vaahtoranta (1994). Since not all states are equally affected by the deterioration of an environmental public good, they distinguish between ecologically vulnerable states, which have a relatively strong interest in pursuing international action through regimes of cooperation, and less vulnerable states, which have a more limited interest in such action. Adding the abatement cost of a country allows us to identify different types of states. States characterised by high vulnerability and low abatement costs are pushers and are expected to favour a more effective regime. States characterised by low vulnerability and high abatement costs are draggers that are expected to want less progress (Sprinz and Vaahtoranta 1994).

The literature on the economics of climate change provides information on benefits from avoiding climate change (vulnerability) and on abatement costs (see, for example, Cline 1992, 2004, Nordhaus 1994, 2001, 2008, Nordhaus and Boyer 2000, Stern 2007), thus allowing us to identify draggers and pushers. In what follows we use estimates from the Regional dynamic Integrated model of Climate and Economy 1999 (RICE-99) runs (Nordhaus and Boyer 2000), complemented with information from RICE-2001 (Nordhaus 2001, 2005), as these provide both country- and region-specific estimates of global warming costs and environmental benefits from abatement. These estimates will allow us to determine the ideal points on the continuum of action on climate change that welfare maximising unitary actors would adopt on the climate issue. Nordhaus, a leading author in the area of climate change economics, employs a discount rate which is neither at the extreme low end, such as Cline's (1992, 2004) and Stern's (2006), nor at the extreme high end of the range (Weitzman 2007, p. 707).[2] Since the discount rate is crucial, and the main driver behind Stern's recommendations for strong and immediate action, we opt for a mainstream position that is in line with empirical observation of economic agents' discounting behaviour as the basis of our explanatory model, rather than an individual author's ethical judgement (Weitzman 2007, Nordhaus 2008). Next we look at the total costs and benefits by country/region of implementing Kyoto. This analysis reveals that the US can be seen as more of a dragger and the EU as more of a pusher with regard to Kyoto. We use estimates that would have been available at the time of making decisions about ratification and implementation of the Kyoto Protocol.

Nordhaus and Boyer's (2000, p. 91) study estimates that the US is characterised by relatively low vulnerability for a 2.5°C increase in temperature, with total damages of 0.45% of GDP in 2100. OECD Europe on the other hand is much more vulnerable, with total damages of 2.83% of GDP. China and Russia are even less vulnerable than the US. In fact Russia would gain from warming, and so would other high income countries on average. India, on the other hand, is very vulnerable with an expected loss in GDP of 4.93%.

We first report the discounted abatement costs and then the discounted net economic impact of implementing the Kyoto Protocol given the economic forecasts of the late 1990s, assuming that the emission targets of the protocol will be extended forever and that the US implements the protocol (Nordhaus and Boyer 2000). These estimates assume trading amongst Annex I countries. Costs and benefits used to calculate the net economic impact are discounted back to 1990 US dollars using the real return on capital over the last few decades, unless otherwise noted. On the basis of these calculations, the discounted cost to the US of implementing Kyoto forever is $325 billion, while OECD Europe gains $5 billon. Eastern Europe, including Russia, gains dramatically at $112 billion, China benefits slightly at $3 billion while high income countries other than the US and OECD Europe face a cost of $24 billion (Nordhaus and Boyer 2000, p. 159).

With regard to the discounted net economic impact (benefits minus costs) of the Kyoto Protocol, Cline (2004, p. 31), using a very low discount rate, finds that the costs to Annex I countries of implementing the protocol are greater than the benefits from reduced warming. Nordhaus and Boyer (2000, p. 160) find that OECD Europe gains overall while the US and the rest of the high income countries lose overall, that is, the discounted costs of abatement exceed the discounted benefits from reduced climate change. This finding suggests that the US had very little reason to ratify and implement the Kyoto Protocol, while the EU would have benefited from implementation by all Annex I countries (Cline 2004, p. 31, Mendelsohn 2004, p. 47), which suggests that the US is more of a dragger and the EU more of a pusher with regard to Kyoto.

More recent estimates (Nordhaus 2001, 2006) suggest an even less conducive environment for the implementation of the Kyoto Protocol. In an updated version, the RICE-2001, Nordhaus (2001, 2006) provides estimates assuming significantly higher long-term factor productivity growth than previously for Europe and the US (Nordhaus 2006). Given these changes in parameters, Nordhaus estimates higher discounted abatement costs for the US and Western Europe for the implementation of the Kyoto Protocol with US participation, amounting to over $2 trillion for the US and about $500 billion for Europe (Nordhaus 2001, p. 1284).[3] The costs of implementing the Kyoto Protocol without the US are much smaller for Europe and other developed countries, but less beneficial for Eastern Europe (Nordhaus 2001, p. 1284). Of course, this comes at the expense of the overall impact of the protocol, which will reduce emissions by only about 1% below the no-policy case (Nordhaus 2001, p. 1283). The very latest estimates suggest, however, that Kyoto has a positive net benefit to the *world* (Nordhaus 2008).

The picture changes when we move beyond Kyoto and consider an optimal trajectory for global greenhouse gases (GHG) reduction. Optimal policies as estimated by RICE-99 lead to much lower total costs to the world than under Kyoto (reduced from $217 billion to $98 billion). However, the discrepancy between OECD Europe, which faces discounted costs of $1 billion, and the US, which faces discounted costs of $15 billion, is still stark (Nordhaus and Boyer 2000, p. 159). Under the optimal policy all but China and Eastern Europe have positive net benefits, even though OECD Europe gets the lion's share (Nordhaus and Boyer 2000, p. 160).

A note of caution is in order when interpreting these figures, as economists generally admit that deriving the above figures is fraught with difficulties. Nevertheless, many students of the economics of climate change hold the view that Kyoto did favour the EU (see, for example, Nordhaus and Boyer 2000, Cline 2004, Mendelsohn 2004).

Domestic politics and climate change: lobbying and campaign contributions

Much of the literature on the politics of global climate change stresses the importance of domestic special interest groups (SIGs), including business and environmental groups (Newell and Patterson 1998, Newell 2000, Boehmer-Christiansen and Kellow 2002, Bryner 2008, Falkner 2008). Particular emphasis is often placed on the role of the business lobby in the US (Boehmer-Christiansen and Kellow 2002, Falkner 2008) and its influence on US climate policy. While much of the literature suggests that SIGs are of importance, relatively little attention is paid to the rational choice literature on interest groups (Fredriksson 1997, Grossman and Helpman 2001, Mueller 2003, Fredriksson *et al.* 2005). There is also insight to be gained from new quantitative studies that demonstrate the influence of interest groups in issue areas other than global warming (Riddel 2003, Binder and Neumayer 2005, Fredriksson *et al.* 2005).

Here we are solely interested in the influence that SIGs exert at the national level of politics, since we are interested in identifying factors that determine the ideal points of states in our international game. Moreover, a number of studies have suggested that it is often easier for interest groups to influence national policy positions rather than intervene on the international level (Newell 2000, p. 102, Boehmer-Christiansen and Kellow 2002).

Some of the formal literature on interest groups has recently found it useful to make a distinction between lobbying on the one hand and campaign contributions on the other. The former is seen as a 'transfer of information by verbal argument' (Grossman and Helpman 2001, p. 104), which might be accompanied by spending on publicity campaigns or to gain access (Lohmann 1995) but not by making quid pro quo payments to any policy makers. The latter part of the literature deals with electoral campaign contributions made to influence policy platforms.

Seen this way, lobbying is essentially a game in which a SIG tries to transmit information about the true state of the world to a policy maker. The assumption is that interest groups have expertise in the areas they are concerned with and that they know, for example, the true implementation costs of a particular policy or can obtain information on the true costs. Policy makers, on the other hand, do not know what the exact costs are; that is, they do not know the true state of the world (Grossman and Helpman 2001, Boehmer-Christiansen and Kellow 2002). It is important to note that this approach assumes that the policy maker is not concerned with gauging the political opinion amongst SIGs; in fact it is assumed that the SIG's bias is known to the policy maker. Thus the model described here is purely concerned with learning about the otherwise unknown true state of the world.[4] The central question, then, is under which conditions a policy maker believes the information that is being communicated by the SIGs. Lobby groups obviously have something to gain from misinforming policy makers in order to get them to make decisions that are favourable to the SIG rather than implement a policy that reflects policy makers' interests (whatever they may be). SIGs have a bias in that they prefer a policy to be weaker or stronger than the one the policy maker would set if s/he knew the true state of the world (Grossman and Helpman 2001).

Grossman and Helpman (2001) discuss a model of lobbying with more than one interest group, which is clearly a realistic scenario for global climate change, where we can identify at least four broad sets of interest groups. First, there are groups such as the Global Climate Coalition representing the fossil fuel lobby and some associated industries, which favour much less abatement than the ideal policy. Then there is a more moderate general industry lobby, which distanced itself in the late 1990s from the position advocated by the fossil fuel lobby (Falkner 2008, p. 138). However, the position of this group is not easily classifiable in terms of bias, although, as it turns out, this does not matter to the conclusions. The nuclear and renewable lobbies and ENGOs are biased towards more abatement than the ideal policy (Newell 2000, Boehmer-Christiansen and Kellow 2002, Falkner 2008). It is not entirely clear which position the insurance lobby takes up, since the insurance industry is on the one hand an institutional investor in the stock market, and thus depends on returns from companies that would suffer from stringent emission targets, while on the other hand the re-insurers in particular fear that extreme weather events might lead to big insurance payouts (Newell and Paterson 1998, Paterson 2001, Falkner 2008).

In addition to assuming that multiple SIGs with biases exist, we also assume that the decision maker actually knows the bias of each SIG. Additionally, we also make the assumption that lobbying is costly and that those costs are endogenous; that is, the interest groups themselves determine the amount of money they are willing to spend on lobbying (advertising campaigns, hiring lawyers and lobbyists, etc.) and that these costs are not necessary to convey the information the group wishes to convey.

In the model of lobbying employed here the principal problem for the policy maker is to determine whether the interest group reports information truthfully. If the policy maker does not know the true state of the world, then s/he implements a policy in line with a weighted function based on prior beliefs about the true value of the state of the world, since this is the 'best guess'. The policy maker would, however, like to learn the true state of the world and then implement *their* ideal policy based on this information. SIGs have the necessary information about the true state of the world, but also an incentive to misrepresent the true state of the world in such a way that the policy maker implements a policy that is in favour of the SIG. The policy maker, on the other hand, knows this. If talk is cheap and the SIG bears no costs then it is difficult for the policy maker to learn the true state of the world due to the incentive of the SIG to misrepresent it.[5]

Introducing endogenous costs into the model helps achieve full revelation of the true state of the world in equilibrium, independent of whether there is one lobby group or multiple lobby groups or one or multiple issue dimensions

(Grossman and Helpman 2001, p. 168). The crucial aspect of endogenous lobbying costs is that the SIGs *do not have to* incur them in order to communicate the information; rather, they incur the costs since 'their willingness to bear a large cost of lobbying ... may signal the urgency of their message' (Grossman and Helpman 2001, p. 183). Endogenous costs allow a SIG to reveal the true state of the world by spending 'enough to convince the policymaker that the reality could not be otherwise' (Grossman and Helpman 2001, p. 162). In other words, SIGs 'buy credibility' (Grossman and Helpman 2001, p. 165) by incurring expenses that are not necessary to transmit the information, expenses that they would never be willing to incur if their costs of implementation, for example, were not as high as they claim. The model shows that in equilibrium the SIG will only be willing to incur such high lobbying costs in order to communicate truthful information.

The more costly the policy which the policy maker would implement without full information is to the SIG (that is, to its members), the higher the costs the SIG is willing to incur in order to send a credible signal that reveals the true state of the world to the policy maker. This is so because the policy maker will then implement *their* ideal policy based on the true state of the world rather than on prior beliefs. It is important to note that the policy maker does *not* implement the policy the SIG prefers most. But the policy s/he implements is better for the SIG than the one s/he would have implemented if s/he had not known the true state of the world. If we consider the above estimates for US abatement costs, therefore, the huge amounts of money the fossil fuel lobby and other SIGs committed to lobbying activities make sense in terms of signalling the true state of the world to the policy maker.

According to this model the policy maker thus learns the true state of the world even if only the Global Climate Coalition lobbies. Of course the assumption that all costs are endogenous is somewhat unrealistic. However, it can be demonstrated that a mixture of some fixed costs and endogenous costs can signal the true state of the world in many cases even if not in all (Grossman and Helpman 2001, p. 168). Additionally, Lohmann (1995) has shown that lobbying by multiple interest groups, which individually do not have full information about the state of the world, also leads to full revelation of the state of the world in equilibrium. Taken together, these results suggest that lobbying in itself is not likely to lead to a socially inefficient policy. However, while the policy maker will set a policy that reflects the true state of the world, the precise policy, and its welfare implications, depend on the policy maker's welfare function and to what degree it is based on *their* own policy interests or desire to be re-elected.

Campaign contributions

Unlike in the case of lobbying, we now assume that direct contributions are made to politicians' campaign funds in exchange for influence on policy positions. In the US this money often comes from so-called Political Action

Committees (PACs) and the sums changing hands are quite significant. Riddle (2003, p. 177) suggests that environmental PACs contributed about $650,000 to the Senate elections of 1996 and 1998. Exxon contributed $1.2 million to George W. Bush's first election campaign (Boehmer-Christiansen and Kellow 2002, p. 36). It is generally argued that business SIGs can easily outspend environmental SIGs (Newell 2000).

So what is the effect on government policy? In order to understand the effect of campaign contributions we again follow a model developed by Grossman and Helpman (2001). They assume that two parties (or presidential candidates) in effect campaign on two platforms and that the policies on the ideological platform are fixed but do not converge. Policy on the second platform is pliable and is set in such a way as to maximise their chance of winning the majority of votes. Voters are split into two groups. Strategic voters vote for the party closest to their interests. Impressionable voters, on the other hand, respond to political advertising and are assumed to vote in line with campaign spending, so that the party that spends more on the campaign gains more votes from this category of voters. Note that without impressionable voters SIGs would not be able to influence the platforms of parties. A SIG's aim is to promote its members' interests on the pliable policy platform and to make campaign contributions accordingly (Grossman and Helpman 2001, pp. 320–328).

Given these assumptions, parties now face a trade-off. A party holds some beliefs about the popularity of its fixed position, but it cannot be certain. It also knows that all strategic voters will vote in line with their interests, that is, they vote for the party closest to their ideal point. Impressionable voters, on the other hand, are responsive to campaign contributions, so that if a party can attract more campaign contributions than its rival by altering its platform, it can obtain more votes from this group. However, moving away from a platform that maximises the welfare of the average strategic voter,[6] in order to attract more campaign contributions, alienates some strategic voters. A party will place more weight on the welfare of strategic voters the larger their proportion. If the fraction of strategic voters is small, however, a party will attach more weight to the welfare of SIG members that offer money. As a result we find that '[i]n equilibrium, the parties act as if they were intending to maximise a weighted average of campaign contributions and the aggregate welfare of strategic voters' (Grossman and Helpman 2001, p. 344). This means that the ensuing equilibrium will only reflect all individuals' welfare if all individuals are either strategic voters or organised in a SIG (Aidt 1998). Since this is not usually the case it means that electoral platforms will to some degree favour special interests and the welfare of the members of SIGs at the expense of others. One implication of this model is that if environmental SIGs do not attempt to contribute to campaigns, policy shifts more heavily towards the welfare of other SIGs.

One might argue that with competing SIGs we should not expect too much of an anti-environmental platform. However, while the policy platform in equilibrium gives some weight to members of environmental SIGs that contribute to the campaign, environmental group membership is only a fraction of the citizens they represent. The basic insight from Olson's theory of collective action is that smaller groups, such as business groups, are more likely to be privileged and thus more likely to overcome the free-rider problem than large groups such as environmental interest groups (Olson 1965). In fact Mueller (2003, p. 498) argues that campaign contributions shift platforms towards the preferred policy of the 'best organised' SIGs.

There is also empirical support for the argument that increasing ENGO membership leads to stronger environmental policies. Fredriksson *et al.* (2005) find empirical support for their hypothesis that a higher number of ENGOs in a country leads to higher pollution taxes in the case of leaded petrol. Similarly, Binder and Neumayer (2005) show that an increase in ENGO membership leads to lower air pollution.[7]

Since ENGOs are less likely fully to overcome the collective action problem than business interests (Grossman and Helpman 2001), we have to conclude that policy platforms in equilibrium are not only tilted towards SIGs, rather than the welfare of the average strategic voter, but also towards well organised business SIGs.

It follows from this discussion that, from this domestic rational choice perspective, a change in a government's position on climate change policy and its negotiation stance might be explained, at least in part, by (1) changes in its abatement costs due, for example, to higher than expected emissions from higher than expected growth; (2) changes to its perceived vulnerability due, for example, to a better understanding of the consequences of climate change; (3) renewed lobbying by SIGs in the light of, for example, new costs; (4) the election of a different president or party with a different ideal policy; or (5) the election of a new president or party influenced by campaign contributions. The first and the latter two points might contribute to our understanding of the change in US policy witnessed after 2001.

International

This section introduces a game theoretical model of international negotiations that has at its core the notion that states' interests diverge in line with our discussion of national policy preferences above. While country positions might change somewhat during negotiations, it is generally acknowledged that they are relatively tightly defined prior to the negotiations (Newell 2000) and can thus be treated as fixed for the purpose at hand.

Like all models, game theoretical models are abstract versions of reality that focus on a number of essential aspects of a social situation. Game theory enables us to gain insights into outcomes by analysing strategic interaction between rational, self-interested actors such as states, policy makers and interest groups, given their interests and some set of rules of the game, such as who has a veto in negotiations.

Game theoretic models help us to understand why there is limited progress on climate change. A stable climate is a relatively pure public good. If a state does not make a contribution to safeguarding the climate, it cannot be excluded from the benefits of the provision of this good; at the same time the good is not diminished if states enjoy it (Barrett 2003). While there is significant evidence that some degree of abatement is in fact the optimal policy from a *global* cost–benefit perspective even after discounting (Nordhaus 2008), this by no means ensures that states will act to supply such a public good, because major reductions in emissions are predicted to be costly (Cline 1992; Nordhaus 1994; Nordhaus and Boyer 2000; Nordhaus 2008) and states thus have an incentive to free-ride. This is typical of a Prisoner's Dilemma.

However, in an *infinitely repeated* Prisoner's Dilemma game, cooperation is possible if states employ conditionally cooperative strategies and do not discount the future too heavily (Ward 1996). Under such a strategy, states would cooperate unless another state defects. Recently there have been some suggestions that cooperation might not be full, either because it is not collectively rational to implement the punishment that is required by conditionally cooperative strategies (Barrett 2003) or because of security externalities (Grundig 2006; Vezirgiannidou 2008).

However, the Prisoner's Dilemma has little to say about the dynamics of negotiations between draggers and pushers; that is, states that the benefit–cost analyses of the Kyoto Protocol suggest have divergent ideal points on the climate policy spectrum. To address this problem we introduce a model we have previously employed and shown to be of use in explaining regime effectiveness (Ward *et al.* 2001, Grundig and Ward 2008). This model assumes that states have divergent interests but does not explain where these come from. The discussion of the political economy theories above fills some of this explanatory gap by allowing us to determine the positions of states in an international negotiation. It also enables us to identify the domestic changes that could lead to a change in negotiation positions.

We can now show how these domestically derived state preferences translate into international outcomes. Since negotiations at international level usually take place under consensus rules, the lowest common denominator might prevail if states have divergent interests with regard to emission abatement. On the other hand, powerful states (leaders) which have major stakes in the negotiations might use their political capital in the form of side-payments to compensate reluctant states for accepting an agreement which they would otherwise veto.

The model features a single issue dimension, in this case climate policy, on which states have divergent ideal points (Figure 1). These ideal points either reflect the interests of the state as a unitary actor or can be seen as a state's negotiation position that is influenced by lobbying and campaign contributions. States with ideal points to the left of the status quo (point 0) oppose action on climate change while states to the right desire varying degrees of action. We split states into leaders and veto states. Two of the states, the US

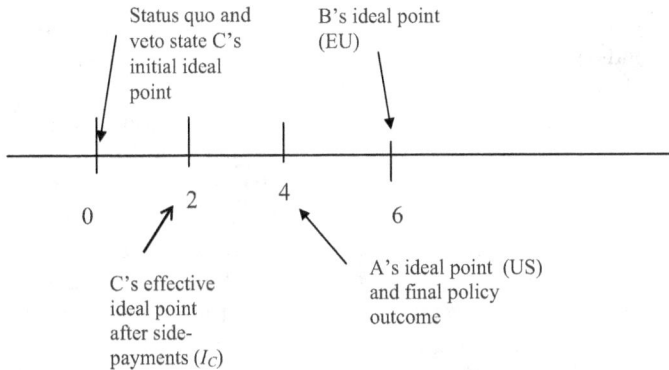

Figure 1. Negotiation model with lead actors on the same side of the status quo and B (EU) with one unit of political capital and A (US) with 2 units.

and the EU, are assumed to be leaders. Only leaders can put together proposals on which veto states vote. A veto state rejects all proposals that are less preferred than the status quo. A veto state's utility depends on how far the proposed agreement is from its ideal point. We assume that there will be at least one veto state that favours the status quo over the ideal point of either leader. We further assume that all players that have to be included in order to achieve an effective regime in the long run are de facto veto powers. Leaders can attempt to influence outcomes in their favour through the use of side-payments, such as offering financial or military assistance, preferential market access or developmental assistance, in order to persuade reluctant states not to veto their proposal. Such side-payments can also be used to persuade veto states to block change. As it turns out, side-payments can be mutually reinforcing or pull in different directions.

We will assume that leaders will not veto any proposal themselves. Of course, we observed in 2001 that leaders like the US *can* renege on agreements after a change in the domestic political configuration. But there are high audience costs associated with such action. If proposals are too progressive, therefore, leaders will usually get veto states to reject the proposal. So our assumption that lead actors can be treated 'as if' they do not have a veto is not as unrealistic as it might appear at first sight.

The model generates a number of different cases (Ward *et al.* 2001), but here we focus on the case that we regard as the closest approximation of climate change negotiations. In this case both leaders' ideal points are to the right of the status quo and the more progressive leader (B), which is the EU, does not have enough side-payments available to move the veto states (exemplified by C) to an effective ideal point I_C at which it is just indifferent between the status quo and the less progressive leader A's (US) ideal point.

If we assume that B has one unit and A two units of political capital, both leaders together have enough political capital to move C to a point at which C would accept a proposal more progressive than A's ideal point. In this case B spends all its political capital (one unit) to move the veto state as far as possible while A spends just enough political capital (also one unit) such that the combined political capital of A and B moves veto state C's effective ideal point I_C up to the point where it will just accept A's ideal point, which results in equilibrium.

The crucial point of the model is that a leader desiring progress has to make sufficient side-payments to *all* veto states in order for them to accept its ideal policy, while the leader that wants less progress only has to make just enough side-payments to obtain its ideal point or, if need be, to make side-payments to *one* randomly chosen veto state in order to stop a proposal that is too progressive. Thus a leader requires multiples of political capital in terms of side-payments or power over a rival to achieve a more progressive policy, while little power is required to block such a policy.

We can draw a number of conclusions from this about the possibilities of achieving an effective future climate regime. First, if the US were more of a pusher, for example due to the policy maker taking a more environmental stance induced by campaign contributions from environmental SIGs representing a large number of citizens, than her resources in terms of side-payments could be fully utilised. Recall that in the example above the US did not use all of its resources and thus stifled progress.

Second, if the US continues to be somewhat of a dragger then investing in EU political capital to make side-payments to veto states does not really help, because in response the US will either exert less of its own resources to achieve progress or, if the EU has enough resources to achieve more progress than the US desires, will use its own resources to hinder progress by making side-payments to a single veto state to block the agreement.

Third, if both the US and the EU desired an effective regime – that is, if both were pushers but did not have enough political capital to pay off veto states – then any movement of veto states towards more progressive ideal points increases the chances of a more effective outcome. This aim could be achieved by environmental SIGs if they managed to overcome the collective action problem better and, for example, made campaign contributions in such states on behalf of its members. However, this model suggests that spending environmental SIG resources on making already very progressive veto states more progressive would have little effect on international outcomes. These resources might be better spent elsewhere.

Future political strategies: participation, information and increasing the returns on green money

Domestic political economy models suggest that the US has a less progressive ideal point on climate policy than the EU, that is, it is more of a dragger while the EU is more of a pusher. This is due not only to the benefit cost ratio of current climate policies but also to a domestic political system that is more conducive than others to special interest politics, with a resulting bias towards business interests. This situation does not bode well for those who desire strong action. Without the US as a more progressive pusher it looks unlikely that an effective regime will emerge post-Kyoto.

Are there strategies, either nationally or internationally, that could improve the situation? Campaign contributions by well-organised SIGs that are opposed to effective action on climate change, the partial collective action failure of citizens affected by climate change, and the policy preferences of President Bush arguably turned the US into more of a dragger than it might otherwise have been. Similar arguments are true for other countries. However, circumstances can change. With regard to the US, President Obama has already indicated that he has policy preferences that are closer to those of the average voter than were Bush's (Bomberg and Super 2009). These preferences should result in a policy more in line with the average voter even if the same lobbying efforts persist and/or if campaign contributions are promised by business SIGs.[8] Most important in this context might well be Obama's ability to bypass business SIGs when it comes to raising campaign funds. Whether the voters who contributed small amounts to his campaign will still be pro-climate change action if this means higher carbon taxes, however, remains to be seen. If in addition the collective action problem of ENGOs could be addressed, then some of the move away from the average voter's welfare could be checked or even turned into a pro environmental bias.

How might a strengthening of pro-environmental groups be achieved?

Participation: solving the collective action problem

ENGOs have the ability to overcome the free-rider problem as political entrepreneurs, at least partially, by offering selective incentives to boost membership (Olson 1965). Without such incentives it might be more difficult to increase membership, especially since the effect of CO_2 pollution is not local and there is a strong temporal asymmetry between the occurrence of pollution and the manifestation of damages (Binder and Neumayer 2005). However, some recent events are encouraging for ENGOs. For one thing, Obama elicited a huge amount of small donations for his election campaign, thus seemingly able to overcome the collective action problem to some degree. The Atheist Bus Campaign in London also offers another beacon of hope in that a clearly latent group donated in excess of £100,000 for an atheist advertising campaign within a couple of days (Sherine 2008). On the international level, organisations such as the World Bank could strengthen civil society in various countries in an attempt to boost the stringency of environmental policy (Binder and Neumeyer 2005, Fredriksson *et al.* 2005). The theoretical model discussed above suggests that improvements in ENGO membership numbers are expected to, *ceteris paribus*, 'green' the negotiation position of states. This is the case because parties and policy makers take the views of SIG members and strategic voters into account when deciding their stance on the pliable policy platform as it relates to the environment. Larger membership might also be of assistance in widening the scope of emissions trading to non-Annex I countries by improving institutional capacity and the strength of civil society to aid verification (Fredriksson *et al.* 2005), thus allowing for a more efficient agreement (see below).

Information: turning impressionable voters into strategic voters

Another way to reduce the influence of SIGs is to reduce the number of impressionable voters. Although rational choice models do not specify why some voters are impressionable and others are not, a rational choice point of view suggests that the cost of obtaining information might be a crucial element. Therefore information campaigns and general engagement of the public might help ENGOs to limit the influence of SIGs with opposed biases by helping to turn more voters into strategic voters. Their views would then be taken into account by parties when positioning themselves on the pliable platform.

Maximising the return on green money

Larger SIG membership increases the weight that a party or candidate attaches to the welfare of that SIG (Grossman and Helpman 2001). Such an increase in membership would also mean that ENGOs have a greater financial clout and thus more influence (Mueller 2003). Research has shown that while environmental PACs make smaller financial contributions to campaigns of environmentally friendly politicians than business PACs, the effectiveness of 'green' money is much higher than that of other contributions. This has been attributed to an endorsement effect (Riddle 2003). Although Riddle's finding is based on a single country (the US) and a limited number of election periods, should this effect be confirmed, it may well open the door for a bigger influence of ENGOs on policy outcomes through campaign contributions. Since most ENGOs pursue multiple spending programmes, the distribution of available funds to different types of activities might have to be re-thought. It is possible that directing more funding from domestic and international ENGOs into campaign contributions in the US might be the most effective way to counter global climate change, since influencing the policy position of the US is crucial for success on the international stage. International ENGOs might want to evaluate their international spending patterns in line with this argument.

Accommodating national self-interest: increases in efficiency and re-balancing of costs between participants

From the perspective adopted here, one reason for American hesitation over the Kyoto Protocol is its negative net economic impact on the US. The benefit-to-cost ratio of the US for implementing the Kyoto Protocol is prohibitively low while the EU's is attractively high, giving the EU some room to manoeuvre. The most important message, though, is that this does not have to be the case. There are two different ways of increasing the US benefits-to-cost ratio, and these can be combined. One obvious solution would be to re-balance the costs between the EU and the US so that both have a benefit-to-cost ratio that is greater than one.[9] The second possibility is to make the agreement overall more efficient by following an 'optimal' policy and thus reduce the costs of abatement and improve benefit-to-costs ratios in general. The most likely source of efficiencies is to allow trading as widely as possible to reduce the marginal costs of abatement (Bang *et al.* 2007). Several authors have derived such optimal policies (for example, Cline 2004, Nordhaus 2008). Nordhaus and Boyer (2000) derive such an optimal policy, that is, a policy that is economically efficient, namely one that is based on the revenue neutral distribution of trading permits to *all* countries.[10] The amount of permits is determined by the externality that GHG emissions create. The benefits-to-cost ratio is greater than one for both the US and the EU under the optimal policy. Although an efficient policy is not necessarily a fair one, the benefits under this optimal policy could be adjusted to allow compensation to any state or region by allocating more or fewer emission permits (Nordhaus and Boyer 2000, p. 123). This allows for a re-balancing of costs between states in line with different principles of fairness, including the possibility of transfers to certain regions (Nordhaus and Boyer 2000, p. 25).

The problem with implementing a more optimal policy lies in the difficulty of widening participation, as this is hampered by factors such as the lack of institutional capacity and monitoring in developing countries needed to make a worldwide permit trading system credible. In this context not only the increase in institutional capacity matters, but also the strengthening of civil society as outlined above.

Conclusion

Many of us might feel that the distribution of costs for reducing climate change should be fair and that it should reflect the degree to which the US and others bear responsibility for the problem of climate change. However, as Morgenthau (1972) reminds us, 'fiat justitia, pereat mundus' ('let there be justice, though the world perish') is not an option for a rational foreign policy, and it will certainly not maximise the world's welfare because getting the US on board is key in terms of both short-term emission reductions and the forging of a long-term agreement that includes lesser developed countries. There are a number of ways in which this can be facilitated, starting with a re-balancing of implementation costs; attempts to make the regime as economically efficient as possible; and increasing the influence of ENGOs through membership drives, information campaigns and re-evaluations of the most effective way to spend 'green' money. However, if increasing ENGO membership is crucial, then it is also crucial that citizens put their money where their mouth is.

Notes

1. It is not clear whether a system can be designed that is not in conflict with WTO rules (Jordan-Korte and Mildner 2008). Even if it were possible to design such a system there would still be a rationale for considering the international level based on the first two points.
2. It has been suggested that Nordhaus takes an extreme view and that Stern or Cline, for example, are more representative. However, it has been acknowledged by Cline (2004) and has been demonstrated for the Stern Review by Nordhaus (2008) that the differences in results are driven by the choice of a low discount rate. Weitzman (2007, p. 707) argues that '[c]oncerning the rate of pure time preference, *Stern* follows a decidedly minority paternalistic view (which, however, includes a handful of distinguished economists) that for social discounting selects the lowest conceivable value ... according to the a priori philosophical principle of treating all generations equally – irrespective of preferences for present over future utility that people seem to exhibit in their everyday savings and investment behaviour'.
3. No other peer-reviewed runs of the RICE model are available to the best of my knowledge.
4. This is one aspect of the lobbying literature; lobbying can also be seen as influencing the policy maker's support function by mobilising the public and thus make the policy maker aware of public opinion (for a short review see Binder and Neumayer 2005, pp. 529–530).
5. The true state of the world can under some condition also be revealed if the SIG does not endure costs (see Grossman and Helpman 2001).
6. Note that it is the average voter rather than the median voter due to the two-dimensional setup with uncertainty (see Grossman and Helpman, pp. 71–72).
7. It is worth noting though that they do not have actual membership levels and thus have to employ a proxy: number of ENGOs per capita (Binder and Neumeyer 2005, pp. 530–531).
8. Opinion polling at the time found that a majority of people was in favour of implementing Kyoto (Lisowski 2002, p. 114).
9. A benefit-to-cost ratio > 1 indicates that the benefits are larger than the costs.
10. It is unlikely, however, that the optimal policy suggested by Nordhaus and Boyer can be fully achieved, as they themselves suggest (Nordhaus and Boyer 2000, p. 123).

References

Aidt, T.S., 1998. Political internalization of economic externalities and environmental policy. *Journal of Public Economics*, 69 (1), 1–16.
Bang, G., *et al.*, 2007. The United States and international climate cooperation: international 'pull' versus domestic 'push'. *Energy Policy*, 35 (2), 1282–1291.
Barrett, S., 2003. *Environment and statecraft*. Oxford: Oxford University Press.
Binder, S. and Neumayer, E., 2005. Environmental pressure group strength and air pollution: an empirical analysis. *Ecological Economics*, 55 (4), 527–538.
Boehmer-Christiansen, S. and Kellow, A., 2002. *International environmental policy. Interests and the failure of the Kyoto Process*. Cheltenham: Edward Elgar Press.
Bomberg, E. and Super, B., 2009. The 2008 US presidential election: Obama and the environment. *Environmental Politics*, 18 (3), 424–430.
Bryner, G., 2008. Failure and opportunity: environmental groups in US climate change policy. *Environmental Politics*, 17 (2), 319–336.
Cline, W., 1992. *The economics of global warming*. Washington, DC: Institute for International Economics.

Cline, W., 2004. Climate change. *In*: B. Lomborg, ed. *Global crises, global solutions.* Cambridge: Cambridge University Press, 13–43.

Conconi, P., 2003. Green lobbies and transboundary pollution in large open economies. *Journal of International Economics*, 59 (2), 399–422.

Falkner, R., 2008. *Business power and conflict in international environmental politics.* Basingstoke: Palgrave MacMillan.

Fredriksson, P.G., 1997. The political economy of pollution taxes in a small open economy. *Journal of Environmental Economics and Management*, 33 (1), 44–58.

Fredriksson, P.G., *et al.*, 2005. Environmentalism, democracy, and pollution control. *Journal of environmental economics and management*, 49 (2), 343–365.

Grossman, G.M. and Helpman, E., 2001. *Special interest politics.* Cambridge, MA: MIT Press.

Grundig, F., 2006. Patterns of international cooperation and the explanatory power of relative gains: an analysis of cooperation on global climate change, ozone depletion, and international trade. *International Studies Quarterly*, 50 (4), 781–801.

Grundig, F. and Ward, H., 2008. Why hegemons do not always get their way. *In: Paper presented at the 104th APSA Annual Convention*, Boston.

Jordan-Korte, K. and Mildner, S., 2008. Climate protection and border tax adjustment [online]. *FACET Analysis* 1. Available from: http://www.aicgs.org/documents/facet/jordan.faceta01.pdf [Accessed 3 April 2009].

Lisowski, M., 2002. Playing the two-level game: US President Bush's decision to repudiate the Kyoto Protocol. *Environmental Politics*, 11 (4), 101–119.

Lohmann, S., 1995. Information, access, and contributions: a signaling model of lobbying. *Public Choice*, 85 (3/4), 267–284.

Mendelsohn, R., 2004. Perspective paper 1.1. *In*: B. Lomborg, ed. *Global crises, global solutions.* Cambridge: Cambridge University Press, 44–48.

Morgenthau, H., 1972. *Politics among nations.* 5th ed. New York: Knopf.

Mueller, D., 2003. Interest groups, campaign contributions, and lobbying. *In*: D. Mueller, ed. *Public choice III.* Cambridge: Cambridge University Press, 472–500.

Newell, P., 2000. *Climate for change: non-state actors and the global politics of the greenhouse.* Cambridge: Cambridge University Press.

Newell, P. and Paterson, M., 1998. A climate for business: global warming, the state and capital. *Review of International Political Economy*, 5 (4), 679–703.

Nordhaus, W., 1994. *Managing the global commons.* Cambridge, MA: MIT Press.

Nordhaus, W., 2001. Global warming economics. *Science*, 294 (9 November), 1283–1284.

Nordhaus, W., 2005. Life after Kyoto: alternative approaches to global warming policy [online]. Manuscript. Available from: http://nordhaus.econ.yale.edu/kyoto_long_2005.pdf [Accessed 21 November 2008].

Nordhaus, W., 2008. *A question of balance.* New Haven, CT: Yale University Press.

Nordhaus, W.D. and Boyer, J., 2000. *Warming the world: economic models of global warming.* Cambridge, MA: MIT Press.

Olson, M., 1965. *The logic of collective action.* Cambridge, MA: Harvard University Press.

Paterson, M., 2001. Risky business: insurance companies in global warming politics. *Global Environmental Politics*, 1 (4), 18–42.

Riddel, M., 2003. Candidate eco-labeling and Senate campaign contributions. *Journal of Environmental Economics and Management*, 45 (2), 177–194.

Sherine, A., 2008. *Atheist bus campaign* [online]. Available from: http://www.justgiving.com/atheistbus [Accessed 21 November 2008].

Sprinz, D. and Vaahtoranta, T., 1994. The interest-based explanation of international environmental policy. *International Organization*, 48 (1), 77–105.

Stern, N., 2007. *The Stern Review on the economics of climate change* [online]. HM Treasury. Available from: http://www.hm-treasury.gov.uk/sternreview_index.htm [Accessed 3 April 2009].

Vezirgiannidou, S., 2008. The Kyoto Agreement and the pursuit of relative gains. *Environmental Politics*, 17 (1), 40–57.

Ward, H., 1996. Game theory and the politics of global warming: the state of play and beyond. *Political Studies*, 44 (5), 850–871.

Ward, H., Grundig, F., and Zorick, E.R., 2001. Marching at the pace of the slowest: a model of international climate-change negotiations. *Political Studies*, 49 (3), 438–461.

Weitzman, M., 2007. A review of the Stern Review on the economics of climate change. *Journal of Economic Literature*, 45 (3), 703–724.

Addressing climate change: a media perspective

Neil T. Gavin

Analysing media coverage can shed light on the politics of climate policy. Conventional notions of how the public sphere represents (or fails to represent) significant social and political issues to citizens are examined. The prominence of global warming stories in the British press is a principal focus. The contours of climate change discourse and the coverage of emissions trading are also touched upon, alongside the political implications that follow. It is then argued that the coverage contributes to a public sphere that is truly international in character but only in a modest way, making political intervention that much more difficult. Climate change policy must take this international public sphere into account if we are to understand how the mitigation of global warming is to be approached and encouraged (and with what limitations).

Introduction

Governments committed to national and international action to reduce levels of CO_2 operate in an environment they wish to shape and control, but one rarely of their choosing. The types of legislative machinery available, the levels of inertia in implementation systems, the structure of party and governmental machineries, and the nature of party-political competition will all condition the success or otherwise of attempts to manage the global warming threat. Some of these factors have static characteristics. Others are more fluid and dynamic. Amongst the latter are the mass media, its coverage of climate change, and its response to legislative initiatives. We now see the 'mediatisation' of politics in contemporary society (Meyer 2002), and in important ways politics is carried on through, and with reference to, the mass media (Mazzoleni and Schulz 1999, Schulz, 2004). For these reasons politicians are especially sensitive to the way in which the media handle issues like climate change, fearing that the media may shape the public's perception of what is important – the media's agenda-setting function – or influence the public's opinions and attitudes not only about the whys and wherefores of the issue but also about government performance. There is a growing literature suggesting that these concerns are reasonably well founded. But even if there was no such impact, it is clear that politicians *think* that the public are influenced, and may act on this premise – a form of 'third person effect' (Herbst 2002, Perloff 2002, Kepplinger 2007).

All these factors make the analysis of the coverage of climate change important. They also draw into view the theoretical concept of the 'public sphere' – that area of public life where common concerns and societal problems are defined and debated, and the domain where issues are aired and, as a result, public opinion is formed (Habermas 1989). The media are deeply involved here, and there is general agreement that the performance of the media is important for a mature and rational political response to the problems besetting modern societies. Although the concept of the public sphere has tended to be applied to the dynamic of debate at the national level, there is a growing sense that it is also relevant to how we understand politics, policy and debate at the supranational level, notably within the European Union (Trenz 2004, Fossum and Schlesinger 2007).

In what follows we will explore the notion of the public sphere in relation to Britain, as an example of a mature western democracy that sees itself as a leader in climate policy. Section 2 will look at what various theorists have concluded about the vitality and adequacy of debate within the British public sphere, before embarking on an empirical investigation of the general level and quality of coverage of climate change, with a specific examination of reports on the European emissions trading scheme (ETS). This is a prelude to an assessment in the third section of the political implications that follow from what can be regarded as dismal, limited and often misleading coverage for informed debate and for political accountability at national and European levels. This will be done alongside an evaluation of the solutions to this problem, especially those canvassed at a theoretical and hypothetical level. The prognosis here is not encouraging, and there are few immediate or obvious prescriptions offering significant change. Section 4 returns to the notion of the public sphere to suggest that we need a concept that transcends the national and European levels to encompass the international level as well, and it will be argued that we must take this into account if we are to understand how political intervention is to be approached and encouraged. What goes on in this 'international public sphere' is often beyond the control of nationally situated actors, but it will be argued that this has positive as well as negative implications for political action on climate change.

Media context and the problem facing governments

There is currently a heated debate about the quality of mediated politics in Britain. There are two schools of thought, which can be broadly characterised as the 'optimists' and the 'pessimists', each marshalling evidence to support a particular evaluation of the public sphere. Key concepts here are 'media influence', 'journalistic practice' and 'the adequacy of coverage'. Media influence relates to the impact of coverage on the public. This is thought to be contingent and variable, depending on such factors as level of exposure to coverage, the strength of an individual's pre-existing political dispositions, the amount of trust people invest in particular media, and the extent to which citizens are dependent on the media for information, as they may be with regard to foreign affairs, technical issues such as climate change, and other concerns beyond their direct experience (McNair 1995, Street 2001). This variability is less significant, however, when it comes to the well-established power of the media to set the public agenda (McCombs 2005) and, as already noted, politicians consistently act in anticipation of what they perceive to be the media's influence on public attitudes. Likewise, 'journalistic practice' varies considerably depending on the type of media, public service considerations, position in the market, and the level of competitive pressure experienced. Finally, the 'adequacy of coverage' can also vary considerably and is dependent, again, on which media are considered, their position in the market and, importantly, the level of sensationalism or tabloidisation evident in coverage (Barnett 1998). The last is of particular concern here, especially where sensationalist coverage threatens to displace hard news or foreign and international commentary, as this has potentially profound implications for effective political accountability (Gripsrud 2000, Gavin 2007a).

Commentators weave these various elements into an overall judgement on debate within the public sphere. The optimists, for instance, believe that television journalists successfully negotiate a delicate balance between educative and entertaining commentary (Marr 2004), a conclusion with some empirical support (Gavin 2007a). The media system is in reasonably good shape too, some argue. Convergent, competitive media, alongside a vibrant internet, offer greater quantities of information from many more sources than formerly – all of it dissected by 'media-savvy' citizens who are less passive than ever (McNair 2006). The media serve citizens who are less than enthusiastic learners, but meet their needs quite adequately (Graber 2003).

The pessimists, on the other hand, claim that even in broadcast journalism the division between news and comment is eroding (Lloyd 2004) and that relentless political marketing means a fixation with process over substance (Blumler and Gurevitch 1995). The PR culture is omnipresent, the blizzard of 'spin' is clearly evident in political coverage (Davis 2002, Davies 2008, Lewis *et al.* 2008a, 2008b), and 'tabloidisation' of coverage is also a concern (Franklin 1997). As a consequence the media are seen as less than effective in holding power to account, with potentially unfortunate consequences for effective and efficient governance.

This is the general backdrop to political and policy-level engagement with the issue of climate policy. The danger is that the media will not treat these initiatives, their limitations or, indeed, climate change itself, with the seriousness they deserve. As a result, climate change may not be high enough on the agenda to stimulate the sort of public concern that prompts concerted political action. Worse still, the media may remain fixated on sensational health scares, the economy, or salacious crime stories, thereby focusing public attention away from climate change. The concern is that it is through coverage of these *other* issues that government competence will be judged. And even if the media do not actually redirect attention to these 'bread-and-butter' issues or shift attitudes about them (although for evidence that it does, see Gavin and Sanders 2003, Sanders and Gavin 2004), the government may *perceive* that it has this impact and act accordingly (Perloff 2002, Kepplinger 2007). This is particularly problematic when more needs to be done on climate change at the same time as other crises also clamour for attention. Doubly so when strenuous intervention is required that involves direct or financially burdensome initiatives, awkward and intrusive regulatory policies, or higher taxation. Such moves will be perceived as having dangerously punishing political repercussions, as we saw in widespread negative reaction to 'pay-as-you-go' road pricing, a policy that the Blair government flagged but which stimulated an avalanche of critical coverage and culminated in two million people signing an on-line petition against the proposals. Even relatively modest impositions on the public, such as increases in fuel duty or air passenger duty, though not very effective in terms of reducing greenhouse gas emissions, have had the same treatment.

In such circumstances political self-preservation virtually dictates a lack of ambition, a degree of caution or an emphasis on non-environmental issues even though this may be wholly inappropriate given the scale and imminence of the climate threat. The urge to self-preservation will be most keenly felt in particular circumstances, notably when governing parties lack a commanding lead in popularity or when (as now) the public do not see 'the environment' as being an important element in their voting decision. This urge will also be felt most intensively when the economy – a perennial preoccupation of voters – is in poor shape. Sadly, this is the very situation the Labour government finds itself in, as, most probably will the next government, whatever its stripe. In such circumstances there is a danger that ideas or initiatives which represent a potential political threat are likely to be toned down, left to languish or replaced by what some call 'cheap-talk environmentalism' (McLean 2007).

These uncomfortable facts of political life render the contours of global warming coverage if not all-important then certainly highly significant. What, then, can be said about the British coverage? The focus here is on the press, since television coverage of climate change is much more limited and, in any event, has much less space to explore the issues within conventional TV news formats (Hargreaves *et al.* 2003). A recent survey of where 'climate change', 'global warming' or 'carbon emissions' figure in story headlines gives a sense of what is going on. The focus was on the four main broadsheet titles, namely *The Guardian, Times, Independent* and *Telegraph*, plus their Sunday sister papers.

Also included were the *Mail, Express, Mirror* and *Sun*, with their Sunday equivalents (Gavin 2007b). After winnowing out duplicate items, letters, and repeated or non-British editions, the picture that emerges shows a spike in coverage in November 2000, with 83 headline stories published at the time of the EU–USA summit at The Hague. However, this was something of an aberration, and coverage dropped off subsequently. The totals for the subsequent years are illustrated in Figure 1. There were 174 stories in total in 2001, dropping to just over 100 in 2002. Thereafter the number rose to 635 in 2006, with another prominent spike of about 90 stories in November 2006 around the time that Stern reported.

Although these numbers look large, that is only superficially true, and they need placing in perspective. The issue here is not one of prominence, although this is important, but of *relative* prominence. If we do a comparable search for health-related stories, with headlines featuring 'health', 'NHS', 'doctor' or 'nurse', we find that there were just over 420 in November 2000 alone, marginally more in the same month in 2006. So stories on this important bread-and-butter issue in one *month* in 2000 outnumbered the *combined* total output on climate change for the next three years, and the number of stories in one month in 2006 was comparable in quantity to the entire output on climate change in 2005. Furthermore, when climate change coverage is disaggregated we find the vast bulk of it in the broadsheet newspapers – predominantly in the left-liberal titles. The mid-market papers carried only marginally over 150 headline stories in the whole of the period from October 2000 to December 2006, with their counterparts in the high circulation 'redtops' carrying just 72, many of them short. Indeed, the *Sun* and the *News of the World* (*NoW*), the two largest selling titles, carried only 18.

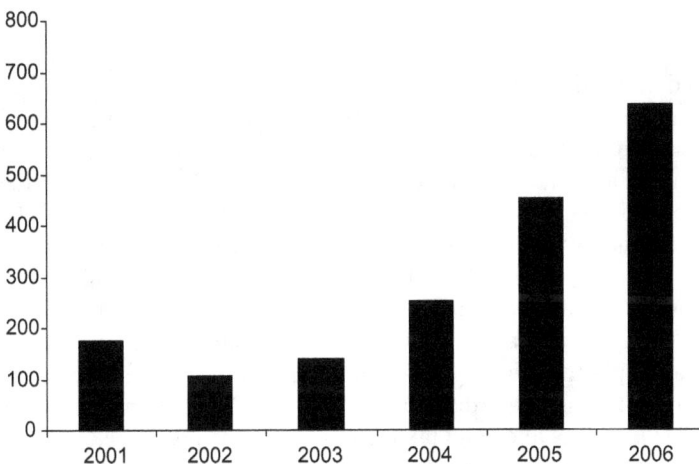

Figure 1. Year-on-year totals of headline coverage of climate change in British national press, January 2001–December 2006.

The coverage, then, was lamentably thin in comparative terms and far outweighed by health stories on issues like postcode lotteries in cancer care, super-bugs and hospital closures, as well as by stories on issues such as the economy and crime. Furthermore, the stance taken by some important titles was quite ambivalent. The *Sun* and *NoW*, for instance, offered mixed messages about the seriousness and imminence of climate change, and the *Telegraph* and *Mail* groups carried worryingly persistent climate denial stories (Gavin 2007b), a feature of their coverage that stretched well into 2007 (Gavin 2007e). If we turn to more specific issues, such as Europe's flagship ETS, the picture is even more dismal. An exhaustive search for stories with headlines containing 'emissions trading' or its synonyms, revealed that over the six years from late 2000 to the end of 2006 there were only 23 headline stories across the entire newspaper output. And these were exclusively in the broadsheet press, overwhelmingly on the specialist 'business', 'finance' or 'City' pages, and predominantly in the *Guardian* and the *Independent* (Gavin 2007d).

Political implications and possible solutions

The limited coverage is unlikely to have convinced readers that climate change is a serious problem warranting immediate, decisive and potentially costly action. This provides limited discursive space for high-profile and ambitious policy interventions, especially those involving the imposition of intrusive, awkward, interventionist or financially costly (and, therefore, potentially dangerous) policies. This is especially the case when resources are stretched, government popularity is on the wane or, more pressing, non-climate-related issues force the government to direct expenditure or invest its political capital and energy elsewhere. In such circumstances, taking the line of least resistance (or the line of least expenditure) is not only expedient but a matter of political survival.

This helps to account for some aspects of current government policy: an unabashed faith that technological innovation will help us manage global warming (Department of Transport 2007); targets that extend into a future that the current political generation will probably not live to see, quite literally; and measures, like the Renewables Obligation, the Climate Change Levy and the European ETS, that only impact on the public *indirectly* and in relation to which, consequently, political responsibility is obscured. The mediated context also helps account for the fact that green taxation has actually fallen as a proportion of all taxes (Environmental Audit Committee 2008). It helps us understand why there is a very modest financial commitment to domestic wind turbine and photovoltaic (PV) installation programmes, and to home insulation upgrading (Guardian 2007, Observer 2008). It may explain much else besides: fuel duty and road tax increases that are postponed in response to short-term conditions; government enthusiasm for road pricing *pilot* projects rather than the real thing; the opening of the hard-shoulder to cars to reduce traffic congestion when reducing road traffic is more appropriate, given the known propensity for greater capacity to lead almost inexorably to greater use; taxes on aviation that are increased, but only modestly; voluntary agreements with car manufacturers on engine efficiency that are unlikely to see the replacement of the existing, high emitting car fleet any time soon; and building regulations that affect the energy efficiency of new structures but only take full effect in 2016 for new homes and in 2019 for non-domestic buildings.

If we turn to the supranational level, the limited coverage identified in our survey has other political implications. Pre-eminent is the almost total absence of ETS coverage, with the result that there is limited (or even non-existent) effective scrutiny and, hence, an absence of meaningful political accountability (Gavin 2007a, pp. 149–153). It is precisely in this sort of situation that we would anticipate some predictable consequences. For one thing, insufficient illumination of powerful corporate interests. Major corporations have often sought to stall, water down or heavily modify climate policy (Newell 2000), and evidently they are still seeking to do so (Financial Times 2008). Furthermore, where there is insufficient attention to the mechanics of a policy, more than ample space is left for muddle, and even for what look like subsidies to CO_2-producing industries. The shambolic disaster that was the first phase of the European ETS perfectly exemplifies this (Bell 2006, BBC 2007, Environmental Audit Committee 2007, Jarman 2007, Open Europe 2007). Much, therefore, remains to be done.

But identifying a problem does not necessarily mean that there are straightforward or readily implemented solutions. Even if the British government wanted to push climate change further up the media agenda, for instance, it may not be in a position to shape the debate that takes place. Governments may wish to affect the coverage of political issues, but are often 'fire-fighting' rather than leading the agenda. Indeed, it is clear that that even in areas central to government activity – such as the economy where, until recently, its record has been fairly solid – the ability of governments to control the debate or influence the media agenda is often heavily circumscribed (Gavin 2007a, 2007d). Governments are not always (and perhaps even not often) able to manage the discursive environment they inhabit. Nevertheless they have to manage *within* it. Furthermore, it is not as if mainstream British non-government organisations (NGOs) or their more militant counterparts will find it easy to force the issue onto the agenda either. Indeed, the direct action approach favoured by some – and perhaps seen in embryo in the 'Plane Stupid' campaign – may backfire and lead to particularly unflattering coverage, as it did with the anti-globalisation and environmental awareness protests in Trafalgar Square in 2000, where the news focused more on the drama of violence and vandalism than on the issues being protested against (Gavin 2007a, pp. 95–118, Gavin forthcoming).

On top of this, the mechanics and technical details of climate change policy do not easily lend themselves to effective communication or elucidation. The media are most comfortable with material that has intrinsic 'newsworthiness', where human interest, personalisation, scandal, surprise, compelling visuals, drama and clarity are the watchwords (Palmer 1998, Harcup and O'Neill 2001). Climate change and policies designed to address it often lack these elements. Stories like Hurricane Katrina, which are dramatic and can be cast as harbingers of things to come in a future of warming oceans, do not happen every day. And even though such natural disasters might stimulate an increased degree of attention to climate change (Boykoff 2007), attributing the phenomenon or any associated damage directly to global warming will not be straightforward.[1] As Simon Lewis, a Royal Society Fellow, put it recently:

> the impact of climate change on human health and mortality is difficult to quantify. There is no comparison group of people not exposed to climate change. Deaths are often due to multiple causes. And while the probability of a particular event occurring under modified climate conditions can be estimated, no single event can be solely attributed to climate change. (Guardian 2008a)

So honest journalists may struggle to connect weather-related events to climate change in a way that will cause citizens to take note or persuade politicians to act.

Added to this difficulty is the fact that a range of themes that emerge from the politics of climate change run against the grain of the neo-liberal economic orthodoxy that seems to pervade the general political scene in Britain as elsewhere. This orthodoxy is still evident in some sectors of the press, and may also be a factor in the Treasury's less than enthusiastic engagement with climate change issues (Carter 2008). The notion that the market's thirst for conspicuous and disposable consumption may be as much a part of the problem as a part of the solution runs against this grain, as does the idea that unbridled personal freedom in car use and aviation transport may not be sacrosanct. Equally contentious from this perspective is the idea that coping with climate change may require market-unfriendly initiatives such as regulation of individual behaviour, regulation of the market or an increased role for the state at local and national level, and beyond. And heaven forbid that we should tax people to incentivise them to change their attitudes or behaviour. Furthermore, if such initiatives cause increased prices, this will be seen as a return to that old bugbear of the Right, inflation.

It is no surprise, then, and may be indicative, that right-wing commentators are conspicuous in their claims that climate change is exaggerated (Booker and North 2007, Lawson 2008), nor that their ideas are eagerly puffed in the right-wing press (Sunday Telegraph 2007, Telegraph 2008). There are also other signs of neo-conservative media mobilisation such as the television documentary 'The Great Global Warming Swindle' – a broadcast used by elements in the press to challenge the veracity of global warming and its anthropogenic roots (Gavin 2007c). Such coverage obviously impedes prominent or balanced treatment of climate change and the politics and mechanics of mitigation.

As for the European Union, a range of authors have already suggested that there needs to be much more coverage of its activities for there to emerge a dynamic European public sphere and therefore meaningful accountability or democracy at this level (Meyer 1999, Peter *et al.* 2003, Trenz 2004, Meyer, 2005). Worryingly, affairs at the heart of Europe are often just not intrinsically newsworthy, as 'the EU policy making processes are complex, arcane, slow-moving, bureaucratic, dispersed, detailed and, in some areas, secretive' so that journalists struggle to make them interesting to, and accessible for, the public (Gavin 2007a, p. 155). With these limitations in mind, we can anticipate equally limited coverage of Phases II and III of the ETS, with a high probability of negative consequences for accountability, policy effectiveness and efficiency, and the successful confrontation of powerful vested interests.

Global climate, global solutions, domestic media

It is not enough simply to raise the profile of climate change in any one, isolated European member state. For maximum leverage to be exerted on the interlocking, hierarchically structured and multi-agency negotiating mechanisms at the heart of Europe, theorists of the 'European public sphere' suggest that we need consistent, parallel and thematically connected coverage that embodies a degree of agenda convergence *across* a range of European countries (Trenz 2004). Only under conditions of concerted media coverage will the minds of politicians and EU officials be concentrated, and coordinated action forced. Yet the notion that we have a developed or developing European public sphere is disputed by many commentators, and with considerable plausibility (Schlesinger 1997, 1999, Peter and De Vreese 2004, Eriksen 2005, Meyer 2005, Fossum and Schlesinger 2007). With regard to emissions trading, the development of an EU public sphere seems rather unlikely because a debate across a wide range of constituent member states is needed for this to become a reality, and this is not occurring. It is certainly not the case in Britain, as we have seen, and even if it were it would also need to encompass the emerging EU democracies as well.

This conception of a supranational dynamic situated in domestic media conditions also has substantive and theoretical relevance beyond Europe. The key element here is the 'global' in 'global warming'. Too narrow a focus on domestic or supranational mediation is in danger of blinding us to the *international* dimensions of coverage. As Al Gore eloquently put it in Bali in 2007, 'we are one people, on one planet, we have one future, one destiny, we must pursue it together'. That destiny has to be pursued at national, supranational and international level. But this draws into the frame the performance of the media in *all* of the countries that are significant players in the theatre of climate change. And it begs us to consider their comparative performance not only within the EU but also in a much wider international setting. Here we are effectively talking about an 'international public sphere' which is analogous to, but wider in scope and grander in scale than, its European counterpart.

This wider array of domestically situated media can handle developments in any number of disparate ways: prompting politicians to action or vigorous intervention, punishing them for actions they think will offend their audience (Britain's 'green taxes' perhaps being a case in point), or remaining mute, thereby allowing political actors space to pursue rational, technically grounded policies insulated from the sorts of frenetic populism that can lead to damaging short-termism. This latter point is part of an argument occasionally put forward by observers who seek to justify the legitimacy of EU policy processes (Héritier 2003, Mény 2003, Schmidt 2004, Lord and Magnette 2004), but it leaves politicians and officials more room than is appropriate to vacillate, procrastinate and tone down policy, or to simply posture. This would mean as little for accountability internationally as it does for the EU (Gavin 2007a, pp. 149–153). Whatever position the media take, the onus is on the quality of coverage, the power of mediated scrutiny and subsequent levels of account-ability in a very wide range of countries, including the USA, China, India, Brazil and Indonesia. And it will concern the media's take (or 'lack of take') not just on global warming and greenhouse gas production but also on issues such as deforestation.

Ultimately this dynamic has serious implications for what can be achieved by way of climate change control, for it draws into play a range of potent forces and complex processes that have mediated, as well as political and economic, dimensions. For instance, Britain and the EU may be acting to establish a raft of policies that set targets, regulate businesses and establish mechanisms for tackling climate change but, rather obviously, they do not do so in an economic or political vacuum, internationally. Powerfully situated economic interests, and no doubt many political actors within the EU, are aware that policies with substantial economic costs that are imposed unilaterally may hand a competitive advantage to businesses in less conscientious countries. Businesses may simply decamp from Europe altogether, for as one recent news story on the EU's raft of developing policies noted,

> Bowing to pressure from energy-intensive industries such as the steel and cement sectors, which have threatened to pull out of Europe if the measures cripple their competitiveness, the commission held out the prospect of free pollution permits under the carbon trading scheme. This came amid warnings that it would be 'economic suicide' to penalise the sectors too heavily. (*Guardian*, 24 January 2008)

The issue is also reflected in controversy over the idea that the EU should impose climate change obligations on companies trading into the EU from countries that do not meet its exacting standards.

This state of affairs is a consequence of globalisation and the fluidity of capital investment and production. But it prompts some probing questions. How far, and how prominently, will domestic media flag up the fact that Europe is offering free permits to big polluters to prevent them from shipping out? How would the media handle the issue in the countries to which such businesses relocate – would they criticise, applaud or remain mute? And what will happen if businesses that trade into Europe are forced to comply with EU emissions capping regulation, with attendant price, sales and profit implications? Will this be portrayed – as it has been in America – as protectionism?[2] And if protectionism is a peg to hang stories on, what political consequences follow? The dynamics here have a bearing on whether we get effective action on climate change or see the unfolding of an international 'tragedy of the commons' in which all countries would like to see something done but few would be first to incur the costs or take the pain (McLean 2007). The likelihood of this happening is at least partly contingent on the media, and this makes the quality of their performance and the maturity of their address (both of which are likely to vary markedly from country to country) a central issue in controlling carbon emissions. Little of this is in the direct control of political actors in any one nation state.

At the same time, however, the international dimension of the development of broader and more effective carbon emissions measures still has some positive features as far as mediation is concerned. We may not as yet have an international public sphere, but we do have the makings of an internationalised discourse. This is scientific rather than political, revolving around the UN Intergovernmental Panel on Climate Change (IPCC). There are a number of aspects of this that are important in media terms. First, IPCC deliberations are regularised and seem likely to continue, so they are easily diarised and reported. Second, if scientists paint an ever-clearer picture with catastrophic and dramatic overtones (Risbey 2008), the issue is likely to get more extensive coverage. And while the IPCC process has political input, the resultant reports are informed and balanced by the science and have to be the product of consensus, which give them a credibility they might otherwise lack. All of this ensures that their conclusions are relayed around the world.

Furthermore, the process itself and how it affects domestic politics is beyond the control of any one nationally situated actor or set of actors – a case in point being the Bush administration's grudging acceptance of the 2007 IPCC report. This is significant because for the best part of a decade the Bush administration and its allies have fought a vigorous and often successful campaign to mute, sideline or contradict the idea that global warming is a reality and that humankind bears some responsibility (Gelbspan 1998, 2004, Nakajima 2001, McCright and Dunlap 2003, Monbiot 2006). But this tack is looking more and more shaky the clearer the consensus on climate change becomes. If the next IPCC report is as worrying as some fear – and there are signs that it might be – climate change will almost certainly be forced much higher up the media, public and political agendas than it has been so far.

Conclusions

None of this would concern us as much as it does if reports of global warming turn out to be exaggerated and its impact is negligible, if targets for emissions curbs were set at appropriate levels, if CO_2 emissions were in decline or stabilising, or if policy actions at the EU level or in Britain were effective and timely. But on almost every front there is cause for concern. As one scientist put it recently, global warming is 'stronger than expected and sooner than expected' (*Guardian*, 23 October 2007), with a danger that the climate may tip into a state of irreparable turmoil (Lenton *et al.* 2008). Consequently current CO_2 targets are probably not ambitious enough. Furthermore, emissions from shipping and aviation are higher than anticipated and likely to increase markedly (Guardian 2008b, 2008c). Emissions overall are on the rise and may have been significantly underestimated (National Audit Office 2008).

These developments obviously demand attention, but in the meantime the realities of domestic mediated politics in Britain virtually dictate that governments should tread warily and with circumspection, distance themselves from responsibility for the negative consequences of climate change intervention, keep direct costs to government as low as possible lest budgets become over-stretched and other pressing concerns have to 'give', and avoid direct criticism for obvious failures of omission or commission in policy implementation or integration, particularly – though not exclusively – with regard to taxation. Commentators may be tempted to criticise New Labour for its palpable failures (Guardian 2008d). But this is too glib and easy, since it underplays the stark realities of life in the world of the 'continuous campaign'. The stance taken by New Labour is not a symptom of cowardice but a tacit acknowledgement of the necessities of contemporary mediated politics. This is largely unavoidable, and is the product of the perceived political costs associated with alternative strategies.

We may want more from governments, or from the press and the public. Little may be happening, but this is not the fault of those who govern, or of the media or the public, since it is as much a systemic as a political issue. For example, conditions in Germany starkly contrast with those in Britain. But the German *system* has a set of interrelated and mutually reinforcing elements that have conditioned the pace of response to climate change. The proportional electoral system allows much more space for green politics at every level of government. This in turn helps to sustain a market for newspapers and magazines which relay and reinforce green concerns (Weidner 2008). Concern about climate change also permeates the legislative system at every level, with a high degree of consensus and engagement at apex-leadership level. The German media have embraced the certainties of climate science (Weingart *et al.* 2000), and consequently climate deniers have less space to peddle their propaganda (Grundmann 2007). This is all very different to the circumstances in Britain.

The current lack of progress is not amenable to quick political or economic fixes. One can simply call for change, as critics regularly do, but this is unlikely to happen any time soon. While blaming the government is easier than prescribing ready solutions, ultimately it is fruitless.

The British economy is entering a period of turbulence and perhaps even decline. Meanwhile the main political parties are scrabbling for public support, and citizens are exercised by a range of issues unconnected to climate change.

All of this conditions and limits the government's room for manoeuvre. It also sensitises the political class to media signals. And these signals are quite clear. Relative to other issues, climate change is short on exposure and, consequently, low on the public agenda, with little likelihood of achieving the prominence of health, crime or the economy. Furthermore, the messages conveyed in the coverage of the threat of global warming are mixed. Consequently the public is not concerned enough – perhaps not *scared* enough – to brook intrusive, expensive or inconvenient intervention without this eliciting a nasty political backlash. Compounding the problem is the fact that in many quarters there is still a visceral hostility to any threat that foreshadows a return to the darkness before the neo-liberal dawn, especially where the impertinences of public expenditure, tax hikes or price rises are concerned. Finally, the British media are as pugnacious, critical and populist as ever, which goes some way towards explaining their frenetic mobilisation against road pricing. Unfortunately the government is not in a position to control any of this. And even if it were, sustained intervention in its own backyard is only part of the picture, as climate change demands action at every level – local, national, supranational and international. This brings into the frame media performance beyond the national level and the limited degree of control over this exercised by any one domestic actor or set of actors.

Nevertheless, the situation is not entirely bleak. The ongoing work of the UN IPCC and the publication of successive reports mean that climate change will continue to force itself onto media agendas around the globe. It will do so almost irrespective of domestic circumstances and in ways that are likely to ratchet up the quantity – if not necessarily the quality – of coverage. And weather catastrophes may yet gain media attention and concentrate minds. Herein lies some hope.

Notes

1. http://news.bbc.co.uk/1/hi/sci/tech/7287988.stm [Accessed 1 March 2008].
2. http://news.bbc.co.uk/1/hi/world/europe/7201835.stm [Accessed 1 March 2008].

References

Barnett, S., 1998. Dumbing down or reaching out: is it tabloidisation wot done it? *In*: J. Seaton, ed. *Politics and the media: harlots and prerogatives at the turn of the millennium*. London: Blackwell, 75–90.
BBC, 2007. Emissions trading. BBC Radio 4, 5 June.
Bell, R.G., 2006. What to do about climate change. *Foreign Affairs*, 85 (3), 105–114.
Blumler, J.G. and Gurevitch, M., 1995. *The crisis of public communication*. London: Routledge.
Booker, C. and North, R., 2007. *Scared to death: from BSE to global warming – how scares are costing us the earth*. London: Continuum.
Boykoff, M., 2007. Flogging a dead norm? Newspaper coverage of anthropogenic climate change in the United States and the United Kingdom from 2003 to 2006. *Area*, 39 (4), 470–481.
Carter, N., 2008. Combating climate change in the UK: challenges and obstacles. *Political Quarterly*, 79 (2), 194–205.

Davies, N., 2008. *Flat earth news*. London: Chatto and Windus.

Davis, A., 2002. *Public relations democracy: public relations, politics and the mass media in Britain*. Manchester: Manchester University Press.

Department of Transport, 2007. *Towards a sustainable transport system: supporting economic growth in a low carbon world*. London: Department of Transport.

Environmental Audit Committee, 2007. *The EU emissions trading scheme: lessons for the future*. London: House of Commons.

Environmental Audit Committee, 2008. *The 2007 pre-budget report and comprehensive spending review: an environmental analysis, third report of session 2007–08*. London: House of Commons.

Eriksen, E.O., 2005. An emerging European public sphere. *European Journal of Social Theory*, 8 (3), 341–363.

Financial Times, 2008. Climate change fears after German opt-out. *Financial Times*, 23 September.

Fossum, J.E. and Schlesinger, P., 2007. *The European Union and the public sphere: a communicative space in the making?* London: Routledge.

Franklin, B., 1997. *Newszak and news media*. London: Arnold.

Gavin, N.T., 2007a. *Press and television in British politics: media, money and mediated democracy*. London: Palgrave.

Gavin, N.T., 2007b. Global warming and the British press: the emergence of an issue and its political implications. *In*: *Paper prepared for the* Elections Public Opinion and Parties *conference*, September. Bristol University.

Gavin, N.T., 2007c. The global warming issue in Britain and the politics of documentary journalism. *In*: *Paper prepared for the* PSA Media and Politics Group *conference*, November. University of Staffordshire.

Gavin, N.T., 2007d. The economy and the limits of political interference: the media in Britain. *In*: *Paper prepared for the Public Policy and the Mass Media workshop*, ECPR Joint Sessions, May. Helsinki.

Gavin, N.T., 2007e. Global warming and peak oil in the British media: the limits of policy development. *In*: *Paper prepared for the* Energy Security in Europe *conference*, September. Lund University.

Gavin, N.T., forthcoming. The web and climate change: lessons from Britain? *In*: T. Boyce and J. Lewis, eds. *Media and climate change*. Oxford: Peter Lang.

Gavin, N.T. and Sanders, D., 2003. The press and its influence on British political attitudes under New Labour. *Political Studies*, 51 (3), 573–591.

Gelbspan, R., ed., 1998. *The heat is on*. New York: Basic Books.

Gelbspan, R., 2004. *Boiling point*. New York: Basic Books.

Graber, D., 2003. The media and democracy: beyond myths and stereotypes. *Annual Review of Political Science*, 6, 139–160.

Gripsrud, J., 2000. Tabloidization, popular journalism and democracy. *In*: C. Sparks and J. Tulloch, eds. *Tabloid tales: global debates over media standards*. Oxford: Rowman & Littlefield Publishers, 285–300.

Grundmann, R., 2007. Climate change and knowledge politics. *Environmental Politics*, 16 (3), 414–432.

Guardian, 2007. Labour's plan to abandon renewable energy targets. *Guardian*, 23 October.

Guardian, 2008a. It's time for a body count. *Guardian*, 26 February.

Guardian, 2008b. Shipping boom fuels rising tide of global CO_2 emissions. *Guardian*, 13 February.

Guardian, 2008c. Aircraft numbers may double by 2026. *Guardian*, 8 February.

Guardian, 2008d. Paler shade of green. *Guardian*, 13 March.

Habermas, J., 1989. *The structural transformation of the public sphere: an inquiry into a category of bourgeois society*. Cambridge MA: MIT Press.

Harcup, T. and O'Neill, D., 2001. What is news: Galtung and Ruge revisited. *Journalism Studies*, 2 (2), 261–280.

Hargreaves, I., Lewis, J., and Speers, T., 2003. *Towards a better map: science, the public and the media*. Swindon: Economic and Social Research Council.

Herbst, S., 2002. How state-level policy managers 'read' public opinion. *In*: J. Manza, F.L. Cook, and B.I. Page, eds. *Navigating public opinion: polls, policy and the future of American democracy*. Oxford: Oxford University Press, 171–183.

Héritier, A., 2003. Composite democracy in Europe: the role of transparency and access to information. *Journal of European Public Policy*, 10 (5), 814–833.

Jarman, M., 2007. *Climate change*. London: Pluto.

Kepplinger, H.M., 2007. Reciprocal effects: toward a theory of mass media effects on decision makers. *Harvard International Journal of Press/Politics*, 12 (3), 3–23.

Lawson, N., 2008. *An appeal to reason: a cool look at global warming*. London: Duckworth Overlook.

Lenton, T.M., *et al.*, 2008. Tipping elements in the earth's climate system. *Proceedings of the National Academy of Sciences of the United States of America*, 105 (6), 1786–1793.

Lewis, J., Williams, A., and Franklin, B., 2008a. A compromised fourth estate? UK news journalism, public relations and news sources. *Journalism Studies*, 9 (1), 1–20.

Lewis, J., Williams, A., and Franklin, B., 2008b. Four rumours and an explanation: a political economic account of journalists' changing newsgathering and reporting practices. *Journalism practice*, 2 (1), 27–45.

Lloyd, J., 2004. *What the media are doing to our politics*. London: Constable and Robinson.

Lord, C. and Magnette, P., 2004. E pluribus unum: creative disagreement about legitimacy in the EU. *Journal of Common Market Studies*, 42 (1), 183–202.

Marr, A., 2004. *My trade: a short history of British journalism*. London: Macmillan.

Mazzoleni, G. and Schulz, W., 1999. 'Mediatization' of politics: a challenge for democracy? *Political Communication*, 16 (3), 247–261.

McCombs, M., 2005. A look at agenda-setting: past, present and future. *Journalism Studies*, 6 (4), 543–557.

McCright, A.M. and Dunlap, R.E., 2003. Defeating Kyoto: the conservative movement's impact on U.S. climate change policy. *Social Problems*, 50 (3), 348–373.

McLean, I., 2007. Climate change and UK politics, from Brynle Williams to Sir Nicholas Stern. *In*: *A paper prepared for the 'The Politics of Climate Change' panel*, European Consortium of Political Research *conference*, September. Pisa.

McNair, B., 1995. *An introduction to political communication*. London: Routledge.

McNair, B., 2000. Journalism and democracy: a millennial audit. *Journalism Studies*, 1 (2), 197–211.

McNair, B., 2006. *Cultural chaos: journalism, news and power in a globalised world*. London: Routledge.

Mény, Y., 2003. De la démocratie en Europe: old concepts and new challenges. *Journal of Common Market Studies*, 41 (1), 1–13.

Meyer, C.O., 1999. Political legitimacy and the invisibility of politics: exploring the European Union's communication deficit. *Journal of Common Market Studies*, 37 (4), 617–639.

Meyer, C.O., 2005. The Europeanization of media discourse: a study of quality press coverage of economic policy co-ordination since Amsterdam. *Journal of Common Market Studies*, 43 (1), 121–148.

Meyer, T. with Hinchman, L., 2002. *Media democracy: how the media colonize politics*. Cambridge: Polity Press.

Monbiot, G., 2006. *Heat: how to stop the planet burning*. London: Penguin Books.

Nakajima, N., 2001. Green advertising and green public relations as integration propaganda. *Bulletin of Science, Technology and Society*, 21 (5), 334–348.

National Audit Office, 2008. *UK greenhouse gas emissions: measurement and reporting.* House of Commons: National Audit Office.

Newell, P., 2000. *Climate for change: non-state actors and the global politics of the greenhouse.* Cambridge: Cambridge University Press.

Observer, 2008. UK lags behind on eco energy. *Observer*, 24 February.

Open Europe, 2007. *Europe's dirty secret: why the EU emissions trading scheme isn't working.* London: Open Europe.

Palmer, J., 1998. News values. *In*: A. Briggs and P. Cobley, eds. *The media: an introduction.* Harlow: Addison Wesley Longman, 377–391.

Perloff, R.M., 2002. Third-person effect research 1983–1992: a review and synthesis. *International Journal of Public Opinion Research*, 5 (2), 167–184.

Peter, J. and De Vreese, C.H., 2004. In search of Europe: a cross-national comparative study of the European Union in national television news. *Harvard International Journal of Press/Politics*, 9 (4), 3–24.

Peter, J., Semetko, H., and De Vreese, C., 2003. EU politics on television news: a cross-national comparative study. *European Union Politics*, 4 (3), 305–324.

Risbey, J.S., 2008. The new climate discourse: alarmist or alarming. *Global Environmental Change*, 18 (1), 26–37.

Sanders, D. and Gavin, N.T., 2004. Television news, economic perceptions and political preferences in Britain, 1997–2001. *Journal of Politics*, 66 (4), 1245–1266.

Schulz, W., 2004. Reconstructing mediatization as an analytical concept. *European Journal of Communication*, 19 (1), 87–101.

Schlesinger, P., 1997. From cultural defence to political culture: media, politics and collective identity in the European Union. *Media, Culture and Society*, 19 (3), 369–391.

Schlesinger, P., 1999. Changing spaces of political communication: the case of the European Union. *Political Communication*, 16 (3), 263–279.

Schmidt, V.A., 2004. The European Union: democratic legitimacy in a regional state? *Journal of Common Market Studies*, 42 (5), 975–997.

Street, J., 2001. *Mass Media, Politics and Democracy.* Basingstoke: Palgrave.

Sunday Telegraph, 2007. Climate of scares. *Sunday Telegraph*, 4 November.

Telegraph, 2008. Lord Lawson claims climate change hysteria heralds a 'new age of unreason'. *Telegraph*, 6 April.

Trenz, H., 2004. The democratizing dynamics of a European public sphere. *European Journal of Social Theory*, 7 (1), 5–25.

Weidner, H., 2008. Climate change policy in Germany: capacities and driving forces. A paper prepared for the 'The Politics of Climate Change' Workshop. *In: ECPR Joint Sessions*, April. Rennes.

Weingart, P., Engels, A., and Pansegrau, P., 2000. Risks of communication: discourse on climate change in science, politics and the mass media. *Public Understanding of Science*, 9 (3), 261–283.

Agenda-setting and climate change

Sarah B. Pralle

An agenda-setting perspective can help us understand current climate policy politics by identifying factors that will help the climate change issue rise and stay high on public and governmental agendas. Keeping climate change at the forefront of government decision agendas will be critical in the coming years because climate change is a long-term problem and governments are unlikely to 'solve' the climate crisis with one policy enacted at one point in time. Kingdon's multiple streams model of agenda-setting is used here to explore strategies for keeping the issue of climate change on agendas and moving it up the list of policy priorities.

Introduction

The problem of human-induced climate change was hypothesised in the early 1890s by Swedish scientist Svante Arrhenius, who warned about the possibility of a so-called 'enhanced' greenhouse effect caused by excess carbon dioxide in the atmosphere (Abatzoglou *et al.* 2007). Despite Arrhenius's early warnings, it took another century before the world's political systems began to recognise and respond to the problem. By the early 1990s, the issue of global warming had secured a spot on the international political agenda, as indicated by the signing of the United Nation's Framework Convention on Climate Change by more than 150 countries at the 1992 Earth Summit. Since that time, many nations (particularly those that committed to specific targets and timetables under the 1997 Kyoto Protocol) have taken steps to reduce their greenhouse gas emissions, with varying success (see DiMento and Doughman 2007). From one perspective, then, we can conclude that climate change is squarely on the agendas of the world's affluent democracies. European Union countries have enacted a number of regional and national laws designed to help meet their

Kyoto targets, the Japanese government has explored market-based policies to reduce their carbon emissions, and the United States under the new Obama administration appears to be on the verge of passing legislation mandating caps on carbon emissions.

If the problem of climate change is already on governmental agendas, then why should we examine the issue of agenda-setting and climate change? By asking what political strategies will enable national governments to make deep cuts in greenhouse gas emissions, the assumption underpinning this volume presumes that policymakers in affluent democracies want to do something about climate change. Yet national governments are not acting more aggressively, it is suggested, because they are afraid of suffering excessive political damage.

The agenda-setting literature suggests a somewhat different set of questions and rests on a different set of assumptions. Agenda-setting scholars ask why some policy issues emerge on governmental agendas while others are relatively neglected (Kingdon 1995). Scholars such as Downs (1972), Cobb and Elder (1983), Hilgartner and Bosk (1988), Kingdon (1995) and Baumgartner and Jones (1993) note that public problems rise and fall on public and governmental agendas, often independently of the objective state of a problem. Indeed, some problems are not defined as problems at all, but rather as conditions with which we choose to live (Stone 1988; Kingdon 1995). Problems without readily available and feasible solutions may fail to get on the decision agendas of governmental actors even if they attract public and governmental attention (Kingdon 1995). Other problems may rise up on the agenda only to fade as the public grows 'bored' and turns to other issues, becomes cynical about our ability to solve the problem, or assumes that it has been solved by the government (Downs 1972). Policy information may be ignored for long periods because of the cognitive limitations of policy actors and institutions, only to receive disproportionate attention at a later date (Jones and Baumgartner 2005b). In short, agenda-setting research examines the fates of different public policy issues as they receive more or less public and governmental consideration, and agenda-setting scholars attempt to explain these varying patterns of attention.

The agenda-setting perspective starts with some basic assumptions. First, scholars have identified at least three broad agendas in democratic political systems, although they use different terminology to describe them. For our purposes, the public agenda refers to the set of issues that are most salient to citizens and voters, the governmental agenda consists of the issues that are up for discussion in governmental institutions such as legislatures and executive agencies, and the decision agenda is the narrower set of issues about which governmental officials are poised to make a decision. Non-governmental institutions, such as the media, also have agendas, and these can affect the public and governmental agendas (Hilgartner and Bosk 1988, Kingdon 1995). The second assumption is that each of these agendas has a 'carrying capacity' that limits the number of issues it can handle simultaneously, thus creating competition among issues for a place (Hilgartner and Bosk 1988). Third, it is less helpful to characterise issues as entirely on or off agendas than it is to think of them as occupying points on a continuum on which some issues are highly salient and a top priority, others are less salient, and still others do not register at all. Finally, the agenda-setting literature assumes that highly salient issues are more likely to move onto the decision agendas of governmental institutions. More effort and resources are expected to be directed to solve these problems than other less salient problems, although policy change is not guaranteed even when an issue is highly salient (Cobb and Elder 1983, Kingdon 1995).

An agenda-setting perspective can help us understand current climate policy politics by identifying factors that will help the climate change issue rise up and stay high on the agendas of governmental and non-governmental institutions. As noted, climate change can be considered 'on' the agenda of many democratic countries, but its position on these agendas varies across time and space. It may, for example, be high on a government's agenda after weather-related natural disasters but then fade as politicians turn their attention to other issues. Keeping climate change at the forefront of governmental decision agendas will be critical in the coming years because climate change is a long-term problem, and ongoing scientific and technological advances will continue to shape (and reshape) our understanding of the problem and the feasibility of various solutions. Put differently, governments are unlikely to 'solve' the climate crisis with a single policy enacted at one particular moment. Instead, the problem requires governments to commit to a series of policy measures, with the probability that progressively more stringent targets will have to be enacted and enforced over time. In short, the climate change crisis requires that the issue remains a priority item that is not displaced by economic downturns and other political, economic and social developments.

Two agenda-setting models provide especially useful insights into how policy issues gain saliency and maintain a central place on public and governmental agendas. John Kingdon's (1995) 'streams' model of agenda-setting devotes attention to how problems get noticed and how issues move onto decision agendas, while Rochefort and Cobb's (1994) problem definition framework investigates how problems are strategically framed so as to increase their salience. Other agenda-setting models, such as Baumgartner and Jones' (1993) punctuated equilibrium model (further developed in Jones and Baumgartner 2005b) are useful for understanding patterns of agenda stability and change and for identifying the factors that drive these dynamics. Kingdon (1995), and Rochefort and Cobb (1994), provide a more in-depth look at the particular factors that increase the odds of a problem receiving a lot of attention, gaining in salience, and achieving high agenda status. The next section outlines these models and begins to apply them to the climate change policy area, using the USA as the primary case for purposes of illustration.

Achieving agenda status: problems, policies and politics

John Kingdon (1995) envisions the rise and fall of issues on the agenda as a product of the interplay of three 'streams' or policy processes: problems, policies and politics. These streams operate largely independent of one another, as they tend to have their own rules, 'star' different players, and are subject to different internal dynamics. Nevertheless, at propitious moments (when 'windows of opportunity' open) savvy policy entrepreneurs can help guide the merging of the three streams, and this merging dramatically increases the chances that an issue will receive serious attention by policymakers. Put differently, when a feasible solution is attached to what the public and policymakers perceive as an important public problem, and when political conditions are amenable to change, a policy window opens. Policy entrepreneurs must then seize the opportunity and push for government action.

The problem stream and climate change

Why would policymakers pay serious attention to climate change at some times but not at others? According to Kingdon (1995), problems come to the attention of policymakers via indicators, focusing events and feedback. Indicators can illuminate the scope and severity of a problem through the monitoring of natural (or social) processes, activities and events. Indicators arise through both routine monitoring and special studies. For example, contemporary scientific and political interest in the phenomenon of global warming was sparked in part by US scientist Charles Keeling's decades-long monitoring of atmospheric carbon dioxide (CO_2) levels, which he began measuring in the late 1950s. His measurements produced what is known as the 'Keeling curve', which shows an alarming trend of increasing carbon dioxide emissions over the last half century (Kolbert 2006). Prior to his study, scientists were not certain whether carbon dioxide would accumulate in the atmosphere or be absorbed by the ocean and by the earth's vegetation. Keeling's research indicated that CO_2 was in fact concentrating in the atmosphere and he provided important evidence suggesting that humans were contributing to the problem (Abatzoglou *et al.* 2007).

Policymakers also learn about problems through dramatic focusing events that grab the attention of the public and policymakers alike. Birkland (1998, p. 54) defines focusing events as relatively rare sudden events that 'can be reasonably defined as harmful or revealing the possibility of potentially greater future harms', and 'are concentrated in a particular geographical area or community of interest'. Unlike problems that are revealed through long-term monitoring using statistical methods, focusing events highlight a problem (or problems) in one striking event that the public and policymakers learn about simultaneously. The flooding of New Orleans and other communities on the Gulf Coast of the United States as a result of Hurricane Katrina was a focal event that suggested any number of problems, including inadequate flood-control protection in the area, inept government response protocols and, perhaps for some observers, the problem of warming oceans resulting from climate change.

Finally, policymakers learn about problems through feedback on current policy programmes. Typically this is negative feedback generated by evaluation studies, target groups, bureaucrats or policymakers themselves, who report on what is not working or on the unintended consequences of policies. For example, initial feedback on the European Union's emissions trading programme suggested problems with its design and implementation that reduced the price of carbon credits and therefore partly undermined the goals of the programme (Clayton 2007).

It is important to note here that even with indicators, focusing events and feedback, issues do not come to the attention of policymakers as 'objective' problems whose meaning is established and uncontested. Instead, much debate exists about whether a problem is amenable to government action, what kind of problem it is, the cause and scope of the problem, and other equally vexing issues (Stone 1988, Baumgartner and Jones 1993, Rochefort and Cobb 1994). Political actors, in fact, spend considerable effort framing problems so as to increase (or decrease) attention to them, mobilise (or demobilise) particular policy actors, and direct policymakers toward preferred solutions. Whose definition of a problem takes hold has enormous consequences because it can shape how an issue is handled in the political process.

Rochefort and Cobb (1994) suggest that policy actors will argue about the severity, incidence, novelty, proximity and crisis nature of an issue, as these factors affect an issue's salience and therefore its agenda status. In general an issue's salience will rise to the extent that policy actors can define the problem as unique and extremely serious, with widespread impacts that hit 'close to home' and result in catastrophic consequences. Importantly, policymakers and advocacy groups will also engage in debates over how to categorise a problem (for example, is climate change a moral problem or not?), the cause of it (who or what is to blame?), and how to solve it (e.g. should we use governmental mandates or market mechanisms?). Indeed, the debate over solutions, or what Kingdon (1995) calls the 'policy stream', is a critical part of the agenda-setting process, and is taken up in the next section.

It is important to note that policymakers process the flow of information from indicators, feedback, focusing events and problem definitions in a disproportionate manner. Jones and Baumgartner (2005a, 2005b) show that policymakers often ignore or under-react to problem indicators in the larger environment. However, circumstances may change (for example, indicators may reveal a severe problem, new aspects of the problem become evident, or new framings of the problem emerge) such that policymakers recognise their error, pay disproportionate attention to a problem, and respond in non-incremental ways to it. The result is a pattern of attention that includes long periods of policy stability punctuated by bursts of agenda (and potentially) policy change (Baumgartner and Jones 1993; Jones and Baumgartner 2005a, 2005b).

The policy stream and climate change

In addition to a problem stream, Kingdon (1995) envisions a policy stream in which solutions are being generated by specialists and experts within policy communities and are waiting to be attached to the salient problems of the day. While there are many potential solutions, only a select few are chosen and implemented. Kingdon argues that proposals must pass a threshold test of technical feasibility and congruence with reigning values to be selected. Moreover, solutions must be perceived as staying within budgetary limits. 'Budgetary considerations prevent policy makers and those close to them from seriously considering some alternatives, initiatives, and proposals' (Kingdon 1995, p. 106). While Kingdon does not dwell on the point, it is worth noting that these criteria are subject to change and that political actors will try to shape the public and policymakers' perceptions about them. Even budgetary constraints, which appear 'objective', are subject to varying interpretations.

For our purposes, the most important point that Kingdon and others make about solutions is the need to have one: problems that have no solutions attached to them are less likely to make it onto governmental and decision agendas. The public is also less likely to worry about problems when they feel there is nothing to be done about them (Abbasi 2006, p. 146). As detailed further below, this characteristic is one of the most critical aspects of climate policy politics: for climate change to rise and stay high on agendas, the public and policymakers must be convinced not only that we *should* do something to combat climate change, but that we *can*.

Politics and climate change

A model of agenda-setting would be incomplete without attention to shifting political opportunities. The multiple streams model focuses on three key political factors affecting agendas: the national mood, organised political forces, and administrative or legislative turnover (Kingdon 1995, p. 146). Kingdon assumes that policymakers sense a 'national mood', perhaps via public opinion polls, and that this mood makes it more likely that the government will pay more attention to some problems and solutions than to others (Zahariadis 1999, p. 77). An 'anti-government' mood, for example, might prevent proposals for large-scale government intervention in the economy and society from achieving a prominent place on the decision agenda. Interest groups may contribute to policymakers' understanding of the public's preferences (or at least the preferences of some segments of it) and of how various solutions will affect target groups (thus influencing policymakers' perceptions of solution feasibility). The balance of interest group support and opposition to a policy may shape policymakers' agendas and selection of alternatives (Kingdon 1995, p. 150). Electoral turnover often leads to rather dramatic agenda changes, as new administrations push their pet issues and raise the status of some problems and solutions. In the USA, significant turnover in Congress can have a similar effect.

Windows of opportunity

The likelihood of any issue rising to prominence on the agenda is significantly increased when the problem, policy and politics streams join together. Such windows of opportunity open as a result of activities in the political stream or because a problem is deemed especially pressing. Kingdon argues that some windows are predictable, such as the budgetary and reauthorisation processes in the US Congress, electoral cycles and the like. Other windows are governed by less predictable processes such as focusing events, damning reports, and the emergence of pressing problems (Kingdon 1995, p. 165). Regardless of whether a window opens predictably or randomly, policy entrepreneurs must be ready to seize the moment, for the windows rarely stay open for very long.

The demise of issues

While Kingdon is primarily concerned with how issues rise on agendas, the decline of issues is a similarly important question. As suggested above, a key challenge in climate policy politics will be to keep the issue high on public, governmental, and decision agendas, as it must weather any economic storms or other developments that might weaken the commitment of the public and policymakers to solving it. Decades ago Downs (1972) predicted that attention to environmental issues would gradually decline after an initial period of enthusiasm and high salience. While Downs' predictions did not bear out – some environmental issues have faded, but new ones have taken their place – he did identify important dynamics that may help explain cycles of shifting issue attention.

Downs identifies the public and the media as driving forces behind issue emergence and decline. Public enthusiasm for solving problems helps to get issues on agendas initially, but subsequent cynicism, unwillingness to sacrifice, or lack of understanding may lead to a decline in attention and agenda status. As the costs and difficulty of solving a problem become more evident, the public tends to lose interest. Similarly, if the public believes that large sacrifices are required (in behaviour, for example), then attention to a problem may wane. A rather different cause of issue decline is when the public (mistakenly or not) believes that the government has solved a problem, and therefore feels free to turn their attention elsewhere. Kingdon (1995, p. 103) adds that even people in government may feel that they have solved a problem and thus redirect their efforts elsewhere. Actual failure to solve the problem can have a similar effect, as policymakers grow tired of trying to pass or amend legislation and let the problem move to the back-burner (Kingdon 1995, p. 103). The media, Downs suggests, can exacerbate cyclical patterns of attention, as it is under pressure continually to find new problems and new solutions, or at least new angles on old ones. As a result, media attention to problems is likely to rise and fall over time; the pattern will depend in part on the nature of the issue and real world events related to it.

In the environmental policy area, the most important 'real world' events affecting the place of environmental issues on the agenda are economic events. Put simply, economic problems often move environmental problems and solutions down the list of priorities. Public opinion data from the United States suggests that the state of the economy is frequently a priority concern and that citizens' willingness to sacrifice economic growth for environmental protection decreases as the strength of the economy declines (Guber 2003).

Political strategies for keeping climate change on the agenda

What can agenda-setting theories teach us about the most effective strategies for putting (and keeping) climate change high on the agendas of affluent democracies? Three sets of strategies are examined here, mirroring the categories (problems, policies, and politics) used in Kingdon's streams model. It is assumed that the actors pursuing these strategies would comprise a range of groups and individuals, including environmental advocacy groups, scientists, journalists, agency personnel, legislators, cabinet members, and perhaps even leaders in renewable energy technologies. Together they constitute the 'climate change advocacy coalition' – the sum total of actors who are active in this policy area and have an interest in getting and keeping the issue high on public, governmental and decision agendas (Sabatier and Jenkins-Smith 1993).

The climate change problem: saliency and strategic framing

To keep the problem of climate change on governmental and decision agendas, it must be salient to policymakers. Public interest in an issue is not the only way to generate salience for politicians in a democracy, but it is an important one. Some politicians will prioritise climate change in their campaigns and when in office based simply on their personal concern for the problem. But policies are not enacted by individuals; therefore, we must consider strategies for increasing its salience more generally so that the average policymaker – someone with limited time and many potential problems to address – is willing to pay serious and frequent attention to the problem.

If the public plays an important role in raising an issue's salience for policymakers, then where does it stand on the issue? A majority of the public in affluent democracies believes that global warming is a 'somewhat serious' to 'very serious' problem, and these numbers have been increasing in recent years (Leiserowitz 2007a). However, expressing general concern for a problem is not an accurate measure of issue salience. A fuller picture emerges when individuals are asked whether they 'personally worry' about a problem, as this suggests a more active state of concern. On this score, the results are more mixed. In the USA only about 19% of respondents in a 2006 Pew Research Center poll said they worried a great deal about global warming compared to 47% who admitted that they worry a little or not at all about it, although recent data suggests growing concern (Pew 2006, Gallup 2008). More recent data from the Gallup Organisation paints a more optimistic picture, as it shows that the proportion of US respondents who said they worried a great deal about global warming reached a high of 41% in March 2007, although it dropped to 37% and then 34% in the two subsequent years (Gallup 2008, 2009). Perhaps the most accurate measure of issue salience, though, results from open-ended questions that ask respondents to list the most important problems facing the country. On these questions, global warming fares poorly, as other issues, like the economy, routinely outrank it when people are asked to list top priority issues for the government to address (Leiserowitz 2007a, Gallup 2008).

One conclusion we can reasonably draw from this data is that there is a significant amount of latent public concern about global warming. Policy entrepreneurs within the climate change advocacy coalition might be able to tap into this latent concern and thereby raise the salience of the problem with the public and policymakers. The key question is how to do it, and this question unfortunately elicits no clear and easy answers. More research is needed to uncover the best strategies for communicating to the public and policymakers in ways that would increase the salience of the issue; the recommendations that follow should be read with this in mind.

1. Regularly report key problem indicators in user-friendly terms

Climate change science has no shortage of indicators. We have data confirming a rise in global temperatures, and scientists are accumulating evidence to show that climate change is affecting sea level, precipitation, polar ice cap coverage, migratory patterns of animals, species habitat, the intensity of hurricanes, and other natural processes (Abatzoglou *et al.* 2007). If Kingdon (1995) is correct in claiming that policymakers (and presumably the public as well) learn about problems through indicators, then the climate change advocacy community should choose a few key indicators to highlight in their communication campaigns. Different indicators may have to be selected for different audiences, depending on the effects that most worry them, but clarity of communication and ease of understanding should be a priority. Using metaphors and analogies could prove useful, as they could help simplify complex scientific relationships (see Stone 1988). However, the climate change advocacy community cannot focus on current indicators alone; it must project into the future so the public and policymakers understand what is likely to happen decades from now. The most alarming indicators for climate change are often based on scientific modelling that looks at future global warming scenarios, which suggests that the public and policymakers must trust scientists and their models; otherwise the indicators are unlikely to raise the necessary alarm.

Indicators alone will not cause agenda and policy change, however. As noted, policymakers and policy institutions often ignore information or discount it for long periods of time (Jones and Baumgartner 2005a, 2005b). In other words, policymakers are unlikely to respond proportionately to changes in problem indicators (that is, attention and policy will not keep pace with changes in indicators); instead they will under- and over-respond over time. Climate change activists must therefore remain flexible and innovative in how they package and frame information about global warming in order to overcome the institutional and cognitive limitations of institutions and policymakers.

2. Emphasise scientific consensus and knowledge

Research suggests that people are less concerned about climate change when they think that scientists do not have a very clear understanding of the issue (Wood and Vedlitz 2007). Moreover, they are less likely to support immediate action to combat climate change when scientific uncertainty is introduced into survey questions (Leiserowitz 2007a). Since less than half of US respondents in a recent survey believe there is consensus within the scientific community (Leiserowitz 2007b), this represents an area for improvement. Every message about climate change should reinforce the fact that the scientific community agrees that global warming is happening and that it is the result of human activity. Moreover, climate change communications should reassure the public and policymakers that scientists know a lot about the problem and its impacts. Policy entrepreneurs should communicate that 'the debate is over' and should point out that the detailed climate models developed by scientists have helped them correctly predict climate change trends and impacts. In other words, communications should not only indicate the extent of the problem but should also address the certainty issue. If people trust that scientists themselves agree and are knowledgeable about this issue, they should be less inclined to discount the problem, more disposed toward considering it a priority issue, and more likely to support immediate actions to combat global warming.

3. Emphasise growing public concern

Individuals take their cues not only from scientists but also from each other. An innovative experiment designed by Wood and Vedlitz (2007) tested how social forces shape people's assessment of the seriousness of global warming as a public problem. They found that when individuals perceived their own definition of the problem to be out of synch with the larger community, they changed their assessment in the direction of that made by the community (Wood and Vedlitz 2007, p. 564). In other words, when respondents were told that 80% of the public viewed global warming as a serious problem, they were more likely to be very concerned about the problem than respondents who were told that only 40% of the public believed it to be very serious. This data suggests that advocates of strong global warming policies should capitalise on the growing public concern about the problem by broadcasting the fact that the 'community' (the national community and global community) views global warming as a serious problem. Such information may prompt individuals who display less concern to 'update' their views to be more in line with the majority. On an aggregate scale, as greater numbers of individuals express high levels of concern, the overall salience of the problem should rise and keep it on agendas.

4. Emphasise specific local impacts and personal experience

Problems that are immediate and proximate to people tend to elicit the most concern from citizens (Rochefort and Cobb 1994). Large proportions of citizens around the world appear to believe that global warming represents a critical threat in the next 10 years (Leiserowitz 2007a). However, respondents in developed countries are less convinced than people in developing countries that global warming will directly affect them, their families and their communities (Leiserowitz 2007a, 2007b). In other words, citizens in affluent democracies tend to think that the impacts of global warming will be geographically distant, affecting people in other countries but not necessarily themselves (Leiserowitz 2007b, p. 8). This is likely to decrease the salience of the issue for people in affluent democracies.

What this research suggests is that the problem of climate change must be defined in ways that emphasise local and regional impacts. Because these impacts will differ depending on the geography and vulnerabilities of particular places, messages should be tailored to different geographical audiences so as to 'bring the issue home'. Wood and Vedlitz's research (2007, pp. 560–561) shows that people who report higher levels of personal experience with the problem (such as those who agreed with the statement, 'My life is directly affected by global warming and climate change') expressed higher levels of issue concern. Policy entrepreneurs should enlist credible spokespeople, such as ranchers, farmers, hunters and others who have close ties to the land, to speak about the changes they have observed. These spokespeople can explain to the public how rising temperatures, changing levels of precipitation, shifting migratory patterns and other issues linked to climate change have affected their livelihoods and recreational pursuits. Not only would these strategies illuminate local impacts, they would also put a human face on the problem and may highlight the economic costs of global warming.

The more general lesson is that the public must be made aware of the specific impacts of global warming whether these are close to home or more distant. Wood and Vedlitz (2007) found that when survey respondents were presented with clear evidence of the effects of global warming, such as rising sea levels, melting glaciers and polar ice caps, and increasingly severe storms, they altered their assessment of the severity of the problem.

5. Emphasise human health impacts

Citizens in affluent democracies are more concerned about the potential human health effects of climate change than any other impacts, according to a 2001 GlobeScan survey (cited in Leiserowitz 2007a). Therefore, in addition to emphasising local impacts, climate change policy entrepreneurs should make it clear to the public and policymakers that global warming may lead to higher death rates from heat waves and higher disease rates due to vector-borne diseases, and that greenhouse gases add to air pollution and its associated health problems. It is important to note, however, that key individuals (within government, for example) and important interest groups may be more concerned with the economic costs associated with climate change, for example, or its implications for national security. Messages to these groups should be tailored to tap into their particular concerns in order to increase the saliency of the problem. Climate change policy advocates must also be prepared to repackage the global warming issue by focusing on new dimensions of the issue. Failure to do so is to risk what Hilgartner and Bosk (1988, p. 63) refer to as 'saturation', whereby the public is flooded with redundant messages that lose their dramatic value and decrease attention to the problem.

6. Insert a moral and ethical perspective into the debate

Many public policy issues are defined in moral terms – abortion, gay marriage, stem cell research, the death penalty, war. It is not a coincidence that these issues attract passionate followers to the point where some individuals privilege these issues above all others when it comes to voting for elected officials, donating to interest groups and making personal choices in their lives. Prominent members of the climate change advocacy coalition have recently called for a more explicit moral framing of global warming (Abbasi 2006), based on the many ethical issues that arise in connection with the issue, including questions of unequal responsibility for the problem, distributional issues (e.g. the world's poor and vulnerable will be most affected), and fundamental questions about the viability of life on Earth. Of course, framing the climate change issue in moral terms carries potential risks because moral arguments can be perceived by some as sanctimonious and may arouse a backlash. Moreover, some of the moral aspects of the problem (e.g. distributional impacts) require citizens in developed countries to consider the impacts of global warming on people who are culturally and geographically distant from themselves. As noted above, distant problems tend to elicit less concern than proximate problems.

Despite these potential limitations, a moral frame could inspire citizens to accept policies that require personal sacrifices because they believe important ethical issues are at stake. Moreover, moral arguments have been shown to increase the mobilisation and political activity of some religious groups, such as the US evangelical community. Given that Christian conservatives in the

United States are a powerful voting bloc in their own right, religious mobilisation could reshape the agendas of reluctant politicians. Conservative politicians who are nervous about being associated with the issue would be provided with 'cover' to support – perhaps even champion – climate change policies.

Climate change policies: offering solutions and salvation

Increasing the salience of the climate change issue must go hand-in-hand with developing and 'selling' solutions to the public and policymakers: problems without attached solutions are less apt to rise high on governmental agendas and are unlikely to make it onto decision agendas at all (Kingdon 1995). Solutions also play a role in keeping issues on agendas. If a solution is perceived as too costly, does not fit with prevailing values, or requires too many sacrifices, then the problem to which it is attached may fade from the agenda. The following discussion does not recommend specific solutions to global warming but rather offers guidance on how to frame them based on agenda-setting theories and general observations about environmental politics and policy.

1. Point to existing solutions

Citizens in affluent democracies appear to support immediate action to address global climate change; in one survey, the majority of respondents preferred that government act now even if there are major costs involved (Leiserowitz 2007a, p. 19). However, this data does not tell us whether the public believes that effective solutions to global climate change exist. To prevent the public from losing interest in the issue due to cynicism about whether the problem can be solved, the climate change advocacy community should emphasise the availability and feasibility of solutions. The message should be that we do not have to wait for some future technology to save us but can begin to implement existing solutions today. Of course some technologies remain controversial, others are not yet ready for 'prime time', and many have not been implemented on a scale that can achieve significant reductions in emissions. Nevertheless, technologies to increase energy efficiency, for example, can often be quickly deployed at relatively low costs (Lovins 2005). These solutions can be presented to the public as starting points, and individuals can begin to implement energy-saving strategies in their own lives, which may empower them to take further actions toward lessening their carbon footprints.

2. Frame solutions in terms of energy

The rising cost of energy, especially oil, raised the salience of the energy issue in the USA during 2007–08, prior to its eclipse by the current economic recession. A 2005 Yale University poll on the environment, for example, found that 92%

of Americans think dependence on foreign oil is a serious national problem (Abassi 2006, p. 49). The climate change advocacy community should capitalise on the public's concern about energy supply and price by focusing on transforming our energy infrastructure. This solution can be attached to a number of policy issues, not just climate change. The energy issue has been linked to national security issues (in the USA, as noted, the focus is 'getting off foreign oil') and to the global economic recession.

Focusing on transforming how we generate and use energy, then, might enlist a broader public in supporting solutions that can also help to mitigate climate change. Moreover, the US public exhibits high levels of support for renewable energy technologies and correctly identifies these as being an important solution to global warming (Leiserowitz 2007a, p. 18). In other words, the public is predisposed toward supporting a transition to cleaner forms of energy, the cornerstone of a comprehensive approach to mitigating climate change. Finally, a focus on energy might make the issue more palatable to politicians who are reluctant to make climate change a centrepiece of their campaigns and policy agendas because they fear the electoral consequences of doing so.

A word of caution is in order, however: recent calls for energy independence in the United States have led to increased support for offshore oil drilling and an increase in the production of biofuels, activities that can exacerbate the climate change problem. It is important, therefore, that the climate change community resists efforts to redefine the energy issue in narrow terms that ignore the problem of global warming.

3. Emphasise the costs of doing nothing

Despite their support for renewable energy technologies, 'many Americans appear to believe that solutions to climate change will be costly and painful, and will offer little by way of corresponding benefits' (Abbasi 2006, pp. 147–148). Agenda-setting theories suggest that this belief is bad news for keeping the issue of climate change on the agenda, as people tend to lose interest in an issue when they perceive the costs of solving it to be too high (Downs 1972). Therefore, in addition to emphasising the economic gains associated with green technology (see below), climate change advocates must point out the costs that countries will incur if they do nothing to address climate change. The Stern Review Report clearly states that the costs of inaction will be far greater than the costs of action (Stern 2007), but it appears that the public has not absorbed this message. The climate change advocacy community should advertise costs that states and localities have incurred as a result of floods, hurricanes and the like as a way of illustrating the potential budgetary impact of rising global temperatures.

4. Focus on economic gains associated with green technology

If global warming is framed in part as an energy issue, then it becomes possible for climate change advocates to speak about jobs and other economic opportunities associated with the transition to clean energy. The Apollo Alliance, a US-based coalition of business, environmental, labour and community leaders, is already framing the issue in these terms. Their 2008 report calls for significant investment in clean energy technologies by the US government, with promises that this will reap considerable economic and ecological dividends (Apollo Alliance 2008). The report makes appeals based on patriotic calls for American innovation and leadership, and taps into concerns about energy independence. These appeals may build support for green energy policies because they draw on widely-held shared values. However, climate change advocates must be careful that they do not detract from the fact that the USA must work cooperatively with other nations to solve the climate crisis, and they must acknowledge that some sectors of the economy will suffer losses.

5. Provide regular feedback about policies and progress

Research suggests that large numbers of US citizens believe that the government shares their concerns about climate change and think that policymakers are doing something about the problem (Abbasi 2006, p. 145). In other words, many Americans are mistaken about the USA's position on the Kyoto Protocol and about its domestic efforts to combat global warming. They appear unaware that other affluent democracies have criticised the USA for its failure to commit to emissions reductions (Abbasi 2006, p. 145). Climate change advocates, then, should make it clear to the public that the USA is considered a laggard when it comes to climate change policy. In fact, advocates should routinely compare – perhaps even rank – affluent democracies on their efforts (and success) toward meeting emissions goals. As Kingdon (1995, p. 100) notes, feedback on policies can alert policymakers to problems and keep them on the agenda. Continual feedback on policies and progress may increase attention to the issue and keep pressure on politicians to meet their commitments.

Climate change politics: maintaining the political will

Of the many uncertainties surrounding climate change politics and policy, perhaps the biggest is whether policymakers in affluent democracies will be able to keep attention focused on the problem over the coming decades. Many things could conspire to lower the priority of the issue even for governments that are predisposed to do something about the problem, let alone those that are more reluctant to act. Moreover, global warming policies may be weakened or overturned after they have been enacted.

Perhaps the best strategy for keeping climate change on the governmental and decision agendas is to design climate change policy in ways that encourage future administrations to pay attention to the problem and discourage future efforts to overturn or ignore it. A policy with long-term commitments that reach over several decades, for example, may provide some protection against agenda decline, although it is not a guarantee. A climate change policy that requires policymakers to revisit emissions goals based on new scientific data may help keep the issue on the agenda as well. Beyond elements of policy design, the climate change advocacy community might engage in the following activities as ways of keeping global warming on the agenda.

1. Take advantage of focusing events

Many focusing events associated with climate change – hurricanes, floods, droughts, heat waves and the like – are particularly 'focal' events, meaning that they do serious, visually dramatic and obvious damage (Birkland 1998). Environmentalists can use the increased public and media attention in the wake of such events to get (or keep) global warming on the agenda. These events should be treated as symbols of climate change, as it is difficult to connect definitively any one storm or weather event to global warming. Environmental groups should point out that a warming world will bring more of these events and emphasise the need to prevent such a volatile (and costly) scenario. Since focal events are often unpredictable, the climate change advocacy community must be prepared to seize the opportunity when an event occurs; in the absence of group activity and mobilisation, 'events will gain little more than passing attention' (Birkland 1998, p. 72).

2. Offer 'predigested' policies to overcome gridlock

The issue of climate change mobilises powerful interest groups that do not always see eye to eye on policy. In the United States, federal action on climate change has been hampered in part by aggressive mobilisation by industry and labour organisations against the climate change advocacy coalition (Kamie-niecki 2006). When powerful interest groups are mobilised on both sides of an issue, policy gridlock may result; as Kingdon (1995, p. 151) puts it, 'Much of the time, a balance of organised forces militates against any change at all' (see also Klyza and Sousa 2008). When gridlock reigns, problems fall off decision agendas because no agreement can be reached. However, competing interest groups might recognise a need to amend policies that are not working, particularly if both sides pay a price for the flawed policy. An example might be a cap and trade programme in which the price of carbon permits is set too low. While some industries may benefit from this, others, such as those who own extra permits, could lose and might be willing to negotiate with environmental groups and others to amend the policy. Under such conditions these unlikely allies could negotiate and offer policymakers a compromise proposal to overcome policymakers' inertia and put global warming back on decision agendas (see Bosso 1987).

3. Venue shopping

Another way in which advocacy groups and policy entrepreneurs can attempt to overcome policy gridlock and inertia is to seek new venues in which to press their policy claims. Baumgartner and Jones (1993) argue that the existence of multiple arenas for policy decision-making increases opportunities for agenda and policy change: if advocacy groups are stymied in one venue, they can appeal to another institution and invite it to assert jurisdiction over the issue. This new institution may take an interest in the problem and put it on its agenda; it may accept a different definition of the problem and therefore give advocates a chance to advance a new understanding of the issue; and it might also provide privileged access to one set of actors so that policy can move forward (see Pralle 2003). Venue shopping gives policymakers and advocacy groups an opportunity to keep issues on the agenda of government by shopping among the various governmental institutions and urging them to address the climate change issue even when others are ignoring it. Studies of forest policymaking and pesticides politics in Canada and the USA (among other research) suggest the success of such strategies for environmental advocacy groups (Pralle 2006a, 2006b).

Conclusion

Climate change is a long-term problem that will require sustained political and policy attention over the coming decades. Policymakers, however, are confronted with a host of legitimate policy problems that are competing for their attention. How can the climate change advocacy community keep the issue of climate change on the agendas of the public and policymakers over the long term? This paper has examined strategies for keeping the issue on agendas and for moving it up the list of policy priorities. Specifically, it has offered strategies for defining the problem of climate change in ways that could raise its salience with the public, on the understanding that if the issue is salient with the public, then policymakers will face some pressure to address it. It is also important to frame solutions in ways that garner maximum support and protect against forces, such as cynicism and fatigue, that might cause the public and policymakers to abandon efforts to solve the problem. Finally, the climate change advocacy community must take advantage of political opportunities, such as focusing events and multiple policy venues, to keep global warming on the agenda even when policy is gridlocked or other issues threaten to displace it.

The specific political strategies for raising the salience of the problem are:

- Regularly report key problem indicators in user-friendly terms;
- Emphasise scientific consensus and knowledge;
- Emphasise growing public concern;
- Emphasise specific, local impacts and personal experience;

- Emphasise human health impacts;
- Insert a moral and ethical perspective into the debate.

Strategies for framing solutions include:

- Pointing to existing solutions;
- Framing solutions in terms of energy;
- Emphasising the costs of doing nothing;
- Focusing on economic gains associated with green technology;
- Providing regular feedback about policies and progress.

Finally, strategies for maintaining political will require:

- Taking advantage of focusing events;
- Offering 'predigested' policies to overcome gridlock;
- Venue shopping.

References

Abatzoglou, J., *et al.*, 2007. A primer on global climate change and its likely impacts. *In*: J.F.C. DiMento and P. Doughman, eds. *Climate change: what it means for us, our children, and our grandchildren*. Cambridge, MA: The MIT Press, 11–44.

Abbasi, D.R., 2006. *Americans and climate change: closing the gap between science and action*. New Haven, CT: Yale School of Forestry and Environmental Studies.

Apollo Alliance, 2008. *The new Apollo program: clean energy, good jobs*. San Francisco, CA: Apollo Alliance.

Baumgartner, F. and Jones, B.D., 1993. *Agendas and instability in American politics*. Chicago, IL: University of Chicago.

Birkland, T., 1998. Focusing events, mobilization, and agenda setting. *Journal of Public Policy*, 18 (1), 53–74.

Bosso, C., 1987. *Pesticides and politics: the life-cycle of a public issue*. Pittsburgh, PA: University of Pittsburgh Press.

Clayton, M., 2007. On global warming, what US can learn from Europe. *Christian Science Monitor*, 30 (January), 2.

Cobb, R.W. and Elder, C.D., 1983. *Participation in American politics: the dynamics of agenda-buildings*. Baltimore, PA: John Hopkins Universtiy Press.

DiMento, J.F.C. and Doughman, P., 2007. Climate change: how the world is responding. *In*: J.F.C. DiMento and P. Doughman, eds. *Climate change: what it means for us, our children, and our grandchildren*. Cambridge, MA: The MIT Press, 101–138.

Downs, A., 1972. Up and down with ecology: the 'issue-attention' cycle. *The Public Interest*, 28 (summer), 38–50.

Gallup, 2008. *Gallup's pulse of democracy: environment* [online]. Gallup Organization. Available from: http://www.gallup.com/poll/1615/Environment.aspx [Accessed 6 October 2008].

Gallup, 2009. *Increased number think global warming is 'exaggerated'*, [online] Gallup Organization. Available from: http://www.gallup.com/poll/116590/Increased-Number-Think-Global-Warming-Exaggerated.aspx [Accessed 5 June 2009].

Guber, D., 2003. *The grassroots of a green revolution*. Cambridge, MA: MIT Press.

Hilgartner, S. and Bosk, C., 1988. The rise and fall of social problems: a public arenas model. *American Journal of Sociology*, 94, 53–78.

Jones, B.D. and Baumgartner, F., 2005a. A model of choice for public policy. *Journal of Public Administration Research and Theory*, 15 (3), 325–351.

Jones, B.D. and Baumgartner, F., 2005b. *The politics of attention*. Chicago, IL: University of Chicago Press.

Kamieniecki, S., 2006. *Corporate America and environmental policy: how often does business get its way?* Stanford, CA: Stanford University Press.

Kingdon, J., 1995. *Agendas, alternatives, and public policies*. 2nd ed. New York: Longman.

Klyza, C.M. and Sousa, D., 2008. *American environmental policy, 1990–2006*. Boston, MA: MIT Press.

Kolbert, E., 2006. *Field notes from a catastrophe: man, nature, and climate change*. New York: Bloomsbury Publishers USA.

Leiserowitz, A., 2007a. International public opinion, perception, and understanding of global climate change. *In: United Nations Development Program, Human Development Report 2007/2008*. Fighting climate change: human solidarity in a divided world. UN.

Leiserowitz, A., 2007b. American opinion on global warming: summary [online]. Yale School of Forestry and Environmental Studies. Available from: http://www.environment.yale.edu/news/Research/5310/american-opinions-on-global-warming-summary/ [Accessed 8 August 2008].

Lovins, A., 2005. *Energy end-use efficiency*. Amsterdam: InterAcademy Council.

Pew Research Center for the People and the Press, 2006. *Global warming: a divide on causes and solutions*. Washington, DC: The Pew Research Center for People and the Press.

Pralle, S., 2003. Venue shopping, political strategy, and policy change: the internationalization of Canadian forestry advocacy. *Journal of Public Policy*, September, 233–260.

Pralle, S., 2006a. *Branching out, digging in: environmental advocacy and agenda setting*. Washington, DC: Georgetown University Press.

Pralle, S., 2006b. Timing and sequence in agenda setting and policy change: a comparative study of lawn care pesticide politics in Canada and the U.S. *Journal of European Public Policy*, 13 (7), 987–1005.

Rochefort, D.A. and Cobb, R.W., 1994. *The politics of problem definition*. Lawrence, KS: University Press of Kansas.

Sabatier, P. and Jenkins-Smith, H., 1993. *Policy change and learning: an advocacy coalition approach*. Boulder, CO: Westview Press.

Stern, N., 2007. *The economics of climate change: the Stern review*. Cambridge: Cambridge University Press.

Stone, D., 1988. *Policy paradox and political reason*. Glenview, IL: Scott, Foresman, and Company.

Wood, D. and Vedlitz, A., 2007. Issue definition, information processing, and the politics of global warming. *American Journal of Political Science*, 51 (3), 552–568.

Zahariadis, N., 1999. Ambiguity, time, and multiple streams. *In*: P. Sabatier, ed. *Theories of the policy process*. Boulder, CO: Westview Press, 73–93.

Clearing the air: the contribution of frame analysis to understanding climate policy in the United States

Amy Lynn Fletcher

Frame analysis illuminates the politics of climate change and generates ideas about discursive strategies that can assist national governments to take effective action on climate change. The nature of frame analysis and its links to discourse theory and social constructivist epistemology are discussed, and this framework used to show how climate change politics in the USA under the second Bush Presidency (2001–2008) have been viewed through at least three contrasting frames: scientific scepticism; climate change as a security threat; and climate change as an economic opportunity. The last of these frames, which uses the Apollo metaphor to liken the task of controlling climate change to the effort during the 1960s to put a man on the moon, is especially promising due to the wide appeal of its positive framing of climate policy in terms of technological achievement, industrial transformation and economic opportunity.

There are no facts, just interpretations – Friedrich Nietzsche.

(Danto, 1965)

Introduction

Frame analysis can contribute to effective climate change policy even though, relative to other methods discussed in this volume, it poses unique analytical and normative issues with respect to the provision of public policy advice. Frame analysis is descended from discourse theory, which in turn is based upon a social-constructivist epistemology that rejects the notion of universal truths and is sceptical about such concepts as objectivity, proof and knowledge accumulation. As Demeritt (2002, p. 776) argues, 'constructionists insist that things are not as they seem [and] that what we had once accepted as self-evidently pre-ordained and inevitable is in fact contingent and might conceivably be remade in some other way, if only we would try'. Frame analysis does *not* focus on whether recommendation X is preferable – more efficient, cost-effective, Pareto optimal – to recommendation Y, as do most mainstream public policy tools and approaches. Instead it focuses on how social actors use language – inclusive of rhetoric, metaphors and storylines – to mobilise key stakeholders, attempt to build a broad public consensus around a course of action, and focus sustained media attention on a specific issue.

Frame analysis is especially powerful when applied to policy conflicts in which parties 'see issues, policies and policy situations in different and conflicting ways which embody different systems of belief and related prescriptions for action' (Schön and Rein 1994, p. xviii). These intractable policy controversies (or 'wicked problems') can be especially difficult to resolve via traditional public policy methods (such as cost–benefit analysis) because disputants begin from different worldviews and assumptions, and hence interpret any evidence or facts via competing and even contradictory lenses. Consequently, 'more research cannot solve differences in perceptions of the problem and its possible solutions, and it cannot prevent research and its results from being ambiguous and contested' (Bueren *et al.* 2003, p. 194). Unlike approaches to climate policy that focus on the identification and measurement of facts, probabilities and the 'best solution', frame analysis exposes the role of political language and worldviews in the construction of plausible, meaningful and socially relevant pathways that can enrol a majority of stakeholders and citizens in collective action. A frame analysis grounded in the significance and ambiguity of political language, in contrast to methods that either bracket language as epiphenomena or assume that the relationship between language and reality is stable, widens the scope of what policy analysts can contribute to understanding such a contested and complex issue as climate change.

The following case study develops this argument by analysing the evolution of climate change policy in the United States during the period 2000–2008. It begins by discussing the frame analysis approach and its links to both discourse theory and social-constructivist epistemology. The next section applies the framing method in order to identify the discursive dimensions of US climate change policy. It identifies the argumentative strategies with which political actors seek to contain the climate change issue and build a stakeholder consensus around a dominant frame that either compels or restrains national public policy responses. In particular the article emphasises the discursive confrontations in US climate politics between global warming scepticism, climate change as a security issue, and climate change as an economic opportunity. Data were gathered through document analysis. Written materials analysed include government documents, policy reports and speeches by relevant actors, publications and websites of non-profit organisations, and mass media articles (a complete list of relevant documents is available from the author). The study concludes that frame analysis deepens our understanding of why nation-states respond as they do to various large-scale environmental challenges and enables the identification of pathways by which even intractable policy conflicts might be successfully re-framed towards consensus solutions.

Discourses, frames and policy problems

Discourse theory, from which frame analysis emerges, includes methods and approaches that range from critical discourse analysis and sociolinguistics to conversation analysis and discursive psychology. What holds these various approaches together is a focus on how language builds – rather than mirrors – social reality. Michel Foucault defined discourse as 'a way of talking about or dealing with a phenomenon. By words, institutions, and practices, discourses realise certain assumptions or views and thus construct their object in particular ways' (Ingstad and Whyte 1995, p. 19). Discourse analysts reject the notion that political language is a stable and accurate reflection of an immutable reality. Social problems and policy solutions do not exist 'out there', waiting only to be recognised and then plucked like apples from a tree and subjected to technocratic analysis. Instead, policy problems emerge *from* discursive interactions within a specific time and place: concepts such as poverty, citizenship and justice are contingent and always subject to change. This does not mean, however, that such concepts are trivial or that there is necessarily no baseline for critique or evaluation across different regimes or institutions.

The discourse analytical task is not to measure and compare discrete, quantitative variables but to evaluate why and how a society comes to accept certain courses of action and policies as preferable. The focus is not only on how actors use language to delineate boundaries and enforce a set of beliefs and social norms but also on how they use language within particular institutional and cultural settings to advance their interests. Rather than beginning with the research question, 'what is the optimal solution to the problem of climate change?', the discourse analyst starts with such questions as, 'why does climate change appear as a policy problem in this time and place?', 'whose interests does foregrounding this problem serve?', and 'how do actors arrayed on all sides interpret the purported facts of the problem?' This approach requires close attention to multiple texts, including mass media articles, speeches, legislation, documentaries, photographs and policy documents, as well as to the ways in which texts are distributed within a society (Phillips and Hardy 2002, p. 5). This linguistic turn, sparked in the late 1960s by such theorists as Jacques Lacan, Roland Barthes and Michel Foucault, poses interesting normative and methodological challenges for the policy sciences. The assertion that 'there is no Archimedean vantage point from which to neutrally evaluate [political] claims' (Barker and Willis 2003, p. 203) does not sit easily within a discipline inspired by Aaron Wildavsky's (1979) call to 'speak truth to power'. Environmental policy analysts in particular can be wary of incorporating social constructivism into the methodological tool-kit. As Bird (1987, p. 256) asks, 'if scientific knowledge is thoroughly a social construction, does this mean that all accounts are equally true, that there are no grounds for one as a more accurate representation than another? That seems an absurd position' (see also Nygren and Rikoon 2008).

However, social constructivism and discourse theory underpin several methods that share a focus on the linguistic and ideational aspects of politics but do not require an *a priori* commitment to radical relativism. Frame analysis in particular provides a method for systematic analysis of the interplay between actors, language and policy. Reber and Berger (2005, p. 186) define frames as 'interpretive structures through which individuals organise and make sense of an ambiguous stream of events and issues in the world' (see also Goffman 1974 for an early and influential discussion of frames). Frame analysts, as Creed *et al.* (2002, p. 36) emphasise, focus on how actors in specific social contexts draw upon familiar cultural accounts in order to produce political change at the organisational or societal level. They identify the main goal of frame analysis as 'understanding how certain idea elements are linked together into packages of meaning, potentially encoded into sound bite-like signifiers that stand for those packages of meaning, and deployed in situated discursive activity' (Creed *et al.* 2002, p. 37). Frame analysis generates powerful insights about the salience of metaphors, storylines and rhetoric that can further the search for culturally meaningful and effective public policies. For instance, Pickerill (2008) found that the word 'wild' means something different to Aboriginal communities than it signifies to most environmentalists. In an Aboriginal context the word 'wild' does not attach to positive images of pristine and protected areas but generates intensely negative associations with the *terra nullius* ('land belonging to nobody') legal fiction by which Britain appropriated the continent. Framing Australia as 'wild country' prior to the arrival of Europeans erases Aboriginals from Australian history and inadvertently complicates the search for environmental policies that can reach across diverse cultures.

Frame analysis also provides a lens through which to evaluate the crucial links between language, actors, institutions and power. In her analysis of the securitisation of HIV/AIDS, for example, Johnson (2006, p. 664) argues that 'framing the fight against AIDS in human rights terms ... drew attention to the applicability of international law to HIV/AIDS issues, and the account-ability of governments and transnational actors'. Phillips and Hardy (2002) discuss the emergence of concepts such as toxic waste and endangered species that changed the relationship between business and the environment in the 1970s and influenced the emergence of an anti-capitalism discourse to contest the dominant globalisation discourse deployed by the major industrial powers. Hajer (1993) concludes that the acid rain controversy in Britain encompassed a much larger crisis of industrial society, raising the question of whether industrialism could be revitalised or if advanced capitalism required a new approach to environmental policy-making. Such insights could only emerge within a methodological orientation that takes language seriously and rejects the assumption of a stable link between a word and an object (between the 'signifier' and the 'signified'). These examples also illustrate that parsing the construction of political reality – and policy disputes – does not have to lead to analytical anarchism. As Hajer (1993, p. 44) asserts, 'large groups of dead trees as such are not a social construct; the point is how one makes sense of dead trees. In this respect there are many possible realities'. The key frame analytical task is to understand what stakeholders and citizens *think* about scientific evidence and facts, and how these different interpretations get translated into public policy and political action. Fischer (2003, p. 127) effectively squares the circle of constructivism and meaningful policy analysis via his attention to the way in which scientific truth is better understood as an 'amalgam of technical and social judgments [in which], in some cases, the technical judgments are more decisive than in others, but both technical and social considerations are always involved'.

Frame analysis does not begin with incontrovertible scientific facts about climate change that can best be interpreted in one way, but with investigating climate change as a concept that stakeholders draw upon to construct a variety of policy problems and to create policy communities. It draws attention to the way in which policy actors try to shape public perceptions of climate change by linking resonant metaphors and cultural resources to a set of recommendations for addressing this issue. Frame analysis also foregrounds the linguistic strategies by which some stakeholders seek to remove climate change altogether from public consideration on the basis of purported doubt regarding climate science, while proponents of climate change mitigation use the same science to justify their appeal for large-scale policy responses.

In the USA, climate change represents one of the most intractable of contemporary policy controversies precisely because there is no stable underlying consensus about what the key facts are, while the standpoints from which stakeholders enter the debate differ significantly in terms of fundamental orientations both to the relationship between market and state and to whether climate change can be addressed within a predominantly pro-growth, free market discourse. Beginning with a discussion of US climate policy in the Bush administration (2001–2008), the following section contrasts the securitisation of climate policy with the emergence of an economic opportunity frame for climate change policy.

Framing the politics of US climate policy

Climate change policy during the Bush administration (2001–2008) is a case of 'frame divergence'. Policy debate about global warming and climate change represented an intractable, or wicked, policy problem wherein differences in stakeholders' frames 'exacerbated communication difficulties, polarised parties and escalated strife' (Shmueli *et al.* 2006, p. 2). A risk and information frame dominated the official American policy response, given President Bush's refusal to support the Kyoto Protocol and the administration's emphasis on the uncertainties in climate change estimates and scenarios. In 2001 President Bush stated, 'no one can say with any certainty what constitutes a dangerous level of warming, and therefore what level must be avoided' (White House 2001b). For proponents of immediate and large-scale climate change policy responses, both in the United States and internationally, this scepticism translated as a deliberate policy of inaction that dovetailed with the economic priorities of Bush's major campaign contributors in the fossil fuel/energy sectors. The ensuing domestic policy debate split into climate 'believers' and climate 'sceptics', with each side routinely enrolling credentialled scientists to dispute the facts presented in the mass media by 'the other side'. As the debate wore on, the risk and information frame morphed into a highly negative characterisation frame wherein disputants sought to undermine opponents' legitimacy and to cast doubt upon their motives (Shmueli *et al.* 2006, p. 4). The caustic rhetorical nature of this intractable controversy deflected attention from actions that the Bush administration did take with respect to climate change, as well as from key institutional factors governing US environmental policy-making. For example, the United States Senate, which must ratify and confirm major treaties entered into by any administration, rejected the Kyoto Protocol in a bipartisan 95–0 vote in 2001. US climate policy also reflects the complicated pluralism of the American system, in which power is divided horizontally and vertically and interest groups have multiple entry points into policy-making processes.

The USA did not ignore climate change in this period, though the actions taken fell far short of those sought by proponents of the Kyoto Protocol. The administration committed the USA to an 18% reduction in greenhouse gas emissions by 2012. It increased funding (with congressional approval) for both the Climate Change Science Program and the Climate Change Technology Program. Moreover, and reflecting the Bush administration's preference for bilateral and multilateral treaties and agreements over global initiatives in trade, security and the environment, by 2005 the USA had concluded bilateral climate reduction agreements with 14 countries and regional organisations that were responsible, in total, for approximately 80% of global greenhouse gas emissions. These countries and organisations include Australia, Brazil, the European Union, China and the Russian Federation (Watson 2005). Ultimately, however, the president's insistence that implementation of the Kyoto Protocol would generate economic costs without compensating environmental benefits, his refusal to identify carbon dioxide as a pollutant that could then be regulated under the existing *Clean Air Act*, and his assertion that climate change policy operated within an 'incomplete state of scientific knowledge of the causes of, and solutions to, global climate change' (White House 2001) inevitably led to conflict with those stakeholders who contend that anthropogenic climate change is the most important existential threat to the planet. Mark Lynas, for example, drawing upon the work of Bob Watson, former chair of the Intergovernmental Panel on Climate Change (IPCC), asserts that 'the harsh truth is that the latest science shows that even two degrees [mitigation] is not good enough, never mind four. And since four degrees [warming] would be a catastrophe that many of us, or our children, would not survive, it is surely our absolute duty to do everything in our power to avoid it' (Lynas 2008).

As in all intractable disputes, however, appeals to the authority of science lost any ability to convince the undecided, as prominent 'sceptics' and 'believers' continued to talk past each other and the debate took on the stale character of oxygen re-circulating in an airplane. In the USA, the Kyoto Protocol failed not only as a public policy opposed by powerful stakeholders but also as a frame capable of enrolling enough committed actors to force policy change at the federal level. For most of the signatories to the protocol, such as the European Union, the document represented not just a techno-scientific policy prescription but a worldview encompassing global cooperation and consensus even in the face of uncertainty. In the USA, however, the Bush administration's stance meant that disputes about the science of global warming remained politically viable at the highest levels of government and in the mass media, while the Kyoto Protocol came to represent (for climate sceptics) the potential erosion of American sovereignty and the external imposition of constraints on economic growth. The American refusal to ratify the Kyoto Protocol, despite the fact that the USA is responsible for approximately 22% of global greenhouse gas emissions and is the world's largest cumulative emitter, alongside its largely unilateral decision to invade Iraq in 2003, merged into a general international image of an administration wilfully tacking against global public opinion and policy. The arid and circular debate between sceptics and believers bogged down US climate discussions at both the domestic and international levels, ensuring that the incremental policies supported by the Bush administration were judged by opponents to be falling short of the actions needed to mitigate global warming. Finally, the global position and influence of the USA meant that the Bush administration positions on climate change also contributed to the failure of even signatory countries to translate Kyoto Protocol prescriptions into meaningful global greenhouse gas mitigation. Though the USA was not solely responsible for the failure of the Kyoto Protocol as either a frame or a set of public climate initiatives, Depledge (2006, p. 14) concludes that 'having to battle against such a powerful force – which actively blocks the discussion of issues and maintains such inflexible positions – places a drag on learning that inevitably exacerbates [regime] ossification'.

American climate change policy during the Bush administration remained stuck in a characterisation frame. Ongoing disputes about the facts and science of global warming elided into accusations of bad faith on all sides, while environmental policy became entangled at both the domestic and international levels with highly intransigent worldviews regarding perceived American unilateralism.

As Saarikoski (2006, p. 618) notes, however, 'if actors' beliefs and interests are constituted through discursive practices, new storylines can create new cognitions and hence influence actors' views of their interests and preferences'. The release of the IPCC's Synthesis Report in 2007, in combination with the

Nobel Prize won by former Vice-President Al Gore and the IPCC, opened a policy window in which the intractable controversy between 'belief' and 'scepticism' could potentially be rewritten into a narrative that supported proactive climate change action.

The 11th hour: global warming as a security issue

Ullman (1983, p. 133) defines a national security threat as a drastic event that occurs in a brief span of time, degrades our quality of life and narrows the range of policy choices available. National security, of course, is a key concept in political science with intellectual roots that date to antiquity. By the mid-1990s, however, political and policy scholars began to focus not only on what national security means, and how it can best be achieved in international politics, but also on the process by which an issue is 'securitised' (Buzan *et al.* 1998). Securitisation refers to the construction of a problem or issue as an *existential* threat to the stability of the nation-state, the 'move that takes politics beyond the established rules of the game and frames the issue as either a special kind of politics or above politics ... something [that] overflows the normal political logic of weighing issues against each other' (Buzan *et al.* 1998, pp. 21–23; see also Waever 1999). The securitisation thesis synthesised many of the post-Cold War debates on the relevance of existing national security institutions and strategies, and became a powerful framework for scholars seeking new ways of thinking about the role of the nation-state and the meaning of security in a global system. Mathews (1989, p. 162), for instance, argues in *Foreign Affairs* that 'environmental strains that transcend national borders are already beginning to break down the sacred boundaries of national sovereignty', though HIV/AIDS, not climate change, was the first issue formally to be securitised at both national and global levels: on 11 January 2000 the United Nations Security Council declared HIV/AIDS a threat to peace and security, the first time a non-traditional issue had been accorded this status. Security Council Resolution #1308 followed, asserting that 'the HIV/AIDS pandemic, if unchecked, may pose a risk to stability and security' (UN 2000).

Containing an issue within a national security frame not only directs political attention to certain concepts, such as threat, danger, enemy, attack and defend, but also prioritises the strategies and norms of national defence and intelligence communities. With the securitisation of HIV/AIDS firmly established by mid-decade, stakeholders in climate policy began to parse the language of security in order to see if it held any relevance to effective policy-making on global warming. Britain's Tyndall Centre for Climate Change Research, for instance, released a working paper on the security implications of climate change. It concluded that the links between security and climate change are indirect in that climate catastrophes could exacerbate, but not usually cause, political tensions. By further eroding state legitimacy and capacity, climate change *might* threaten collective economic livelihood, reduce access to fresh water and food, accelerate exposure to new diseases, undermine military cohesion and increase social inequalities (Barnett 2001, p. 4; see also Brown and Crawford 2008).

Despite the difficulty of establishing significant causal links between national security and climate change variables, the security frame does draw upon both a language and a set of resources that can reconcile previously antagonistic stakeholders. In the American context, successfully securitising an issue places it squarely within the sightline of the White House and significantly increases the likelihood of both media attention and substantial financial support. In combination with media representations of catastrophic events, a focus on security can also change the calculus of cost versus benefit in the direction of pre-emptive action. Against the backdrop of the 2005 Katrina hurricane disaster in New Orleans, for example, which cannot be directly attributed to global warming but which did result in powerful media images of humans caught in an environmental disaster, climate change emerged as an especially attractive candidate for securitisation as 'the storm gave Americans a visual image of what climate change ... might mean for the future' (Busby 2007, p. 1).

In addition to prioritising two of the most powerful institutions in the USA – defence and intelligence – a security frame also provides centrist Republicans and Democrats with a platform for cooperation on climate change mitigation legislation. For example, Senator John Warner (Republican-VA) co-sponsored a greenhouse gas (GHG) reduction bill with Senator Joe Lieberman (Democrat-CT) in the 110th Congress (2007/08). Significantly from a framing perspective, Senator Warner, who had previously opposed global warming-related legislation, justified his decision to support S. 3036 primarily in terms of his professional emphasis on national security, asserting that 'the problem of global climate change [fits] squarely within that focus' (Walsh 2008). The securitisation of climate change in the USA gained further momentum in 2007 with the release of a report from the CNA Corporation, an influential non-profit research institute with close ties to the US Navy, which concludes that 'projected climate change poses a serious threat to America's national security. [It has] the potential to disrupt our way of life and to force changes in the way we keep ourselves safe and secure' (CNA Corporation 2007, p. 6). The report recommended becoming a more 'constructive partner' in international climate change mitigation efforts, as well as increasing energy efficiency in Department of Defense procurements and assessing the potential impact of climate change on US global military installations. Following the release of this report, several powerful actors took up the security frame to focus public, political and media attention on global warming. In addition to Senate and Congressional hearings on climate change as a national security threat (2007 and 2008), various academic and non-profit organisations contributed to the construction of the climate security/national security frame, including the Centre for Strategic and International Studies, George C. Marshall Institute, US Army War College and Triangle Institute for Security Studies, Woodrow Wilson International

Centre Environmental Change and Security Program, Council on Foreign Relations, and National Resources Defence Council.

The green economy and economic transformation

In June 2005, Representative Inslee (Democrat-WA) introduced the *New Apollo Energy Project Act* (H.R. 2828), which sought to develop a cap and trade emissions system and to provide $7 billion in loan guarantees for the development of clean coal power plants, and linked American national security with energy independence, promising a 'comprehensive clean energy plan to address jobs, national security and climate change'. Though this bill did not become law, it did signal the emergence of a new frame for US climate change policy. Just as the Manhattan Project metaphor directs attention to the rapid development of the nuclear bomb during wartime, the Apollo Project metaphor resonates with another American technological triumph: its success in putting a man on the moon and safely returning the rocket's crew to earth in 1969. The US Congress authorised the establishment of the National Air and Space Administration (NASA) in the late 1950s as part of America's immediate response to the Soviet launch of the Sputnik satellite in 1957. Deliberately created as a civilian agency, distinct from the Department of Defense, NASA's Apollo Project gathered momentum following President John F. Kennedy's famous speech (25 May 1961) in which he urged American scientists and engineers to put a man on the moon 'before this decade is out' (Kennedy 1961). The *New Apollo Energy Act* linked this national memory with contemporary action on climate change, rewriting the debates over scientific uncertainty and security threats into a storyline that emphasises using the 'talent that put us on the moon to create a new energy future for our nation ... This is a different challenge than in the 1960s, but we are the same people as we were then – the most innovative on earth' (Inslee 2005, p. 1).

The Apollo Project, which in its Cold War context actually had *both* civilian and national security purposes, nevertheless does not carry the same complicated emotional freight as the Manhattan Project. As a metaphor, an Apollo Project for climate change mitigation and adaptation is more elastic in its ability to mean different things to different constituencies. It is also more stable in that the memory that underwrites it was written from the outset as a positive example of peaceful scientific exploration *and* national technological achievement. In combination with the transition from the Bush administration to that of President Barack Obama, who supports enlarged domestic programmes to mitigate climate change and who seeks to assert American leadership on the global warming issue, the Apollo metaphor provides the basis for a frame that can recast a debate often couched in terms of obligations (higher taxes, more regulation, less growth, more guilt) and threats into one that foregrounds the economic and social opportunities inherent in the construction of a 'green economy'. It also has the potential to expand the number and type of supporters for climate action beyond the additional constituencies that mobilised around the national security implications of global warming – although it can enrol those as well via its emphasis on energy independence.

The Apollo Alliance, a national organisation with numerous state and local affiliates, deliberately draws upon the trope of the US space programme – and collective action in the face of a technological challenge – to underwrite its vision of a 'green transformation' of the USA economy. The Alliance seeks to integrate separate stakeholder groups, such as organised labour, energy corporations, community development groups, small business and environmental activists, into a cohesive collective action movement to 'rebuild the American middle-class' and to make a transition to clean and sustainable technologies on a national scale. For example, the Los Angeles Apollo Alliance convened in 2006 is an affiliate member that represents 26 community-based organisations drawn from the labour, environmental and social activist sectors. It links the development of Green Industry (new technologies for clean energy) to the revitalisation of inner-city communities and the provision of quality job opportunities for the working class, asserting that 'Los Angeles can lead the nation by uniting behind a green economy agenda that is based upon the shared values of environmental sustainability and economic prosperity for all' (SCOPE 2008). At the national level, the Alliance's *New Apollo Program for Clean Energy and Good Jobs* (Apollo Alliance 2008), which is based on an estimated generation and investment of $500 billion over 10 years, seeks to build 'America's 21st century clean energy economy' via the creation of five million 'green jobs', the development and deployment of clean energy sources such as wind and solar, and the integration of energy security, climate stability and economic prosperity.

The Apollo metaphor incorporated into, and deployed by, the Apollo Alliance and its national network is more likely than the national security frame to motivate a large-scale American consensus on the need to shift beyond climate science scepticism towards climate-related policies that would significantly mitigate American greenhouse gas emissions. By enveloping climate change into a frame of industrial transformation and economic opportunity, and by drawing upon an intrinsically positive and powerful cultural narrative, the Apollo Alliance opens up multiple pathways into the climate change issue and brings together multiple and varied interests. Its assertion that 'in the spirit of the American frontier, the future we envision is one not of limited horizons, but of expanding possibilities' (Apollo Alliance 2008) also provides an easy rhetorical hook for politicians across the political spectrum by appealing to American vitality and optimism.

Implications of frames for political strategy

The discursive and institutional links between climate change and national security provide an alternative frame that downplays the stalemate between believers and sceptics. It also gives stakeholders across the American political spectrum powerful rhetorical resources to draw upon in mobilising widespread public action on climate change. In his Nobel Prize speech, for example, former Vice President Al Gore asserted, 'we have begun to wage war on the earth ... We and the earth's climate are locked in a relationship familiar to war planners: mutually assured destruction' (Gore 2007). From the Republican side, as Representative Bob Inglis (R-SC) and Arthur Laffer (2008) argue, 'conservatives do not have to agree that humans are causing climate change to recognise a sensible energy solution. All we need to assume is that burning less fossil fuels would be a good thing.' The 'climate war' and 'energy security' metaphors can both underwrite and link American and international initiatives. For instance, the UN Security Council held its first debate on the impact of climate change on peace and security in April 2007. Echoing the language of existential threat, the representative from Papua New Guinea, representing the Pacific Island nations, asserted that 'the impact of climate change on small islands was no less threatening than the dangers guns and bombs posed to large nations' (UN 2007). Soon after this meeting, the UN Deputy Secretary-General, Asha-Rose Migiro, warned of an imminent global catastrophe unless the major powers addressed climate change 'on a war-footing' (UN 2008).

At the domestic level the Manhattan Project metaphor, with its obvious links to a national security frame for climate change action, now circulates throughout the American debate on global warming mitigation and adaptation. Referring to the Manhattan Project, a secret and successful project, in cooperation with Britain, to develop the nuclear bomb before Nazi Germany did, it draws upon a cultural repertoire of familiar stories and images that reinforces America's sense of itself as an innovative and resourceful country. Stakeholders can also use the Manhattan Project metaphor to impart a sense of urgency to the climate change threat. In a presentation to the National Academies' Committee on Science, Engineering and Public Policy, for example, Norris (2008, p. 5) asserts that 'there is a clear need for large-scale, government-led efforts to develop "transformational technologies" such as solar and wind power. It is the technical problems that can most benefit from applying Manhattan Project lessons.' Finally, by securitising climate change, strategists can potentially persuade military/ security stakeholders who do not typically prioritise environmental issues to take climate change seriously.

However, this frame strategy also poses political risks, especially in comparison to the Apollo alternative. While the 'climate security' approach will appeal to many due to its positive connotations with American power, technological ingenuity and full-scale commitment in the face of an existential threat, it also could alienate stakeholders for whom environmental issues such as climate change must remain foremost a case of human security as distinct from national security. The Manhattan Project metaphor is not just a trope, nor is the security frame simply a set of associations; instead both embody a particular approach to an issue in which certain actors, norms and strategies would be more powerful than others. Language and institutions intertwine in a frame analysis, hence the 'crux of the problem is that ... understanding climate change as a security issue risks making it a military rather than a foreign policy problem, and a sovereignty rather than a global commons problem' (Barnett 2001, p. 11).

In contrast the Apollo frame, with its emphasis on industrial transformation and economic opportunity, is more likely to become a master frame (Reber and Berger 2005, p. 186) in the American context. As deployed by the Apollo Alliance, the 'new' Apollo project encompasses resonant associative metaphors such as green jobs, green energy, green communities and green justice, and has the potential to shape the strategies of the opposition and create linkages among a broader number of interest groups. Drawing upon a powerful and positive narrative, the economic opportunity frame opens up numerous entry points into climate change policy, ranging from mitigation as a test of American leadership to energy independence as a national security imperative to the green economy as a pathway to economic growth and social justice. The explicit Apollo Alliance emphasis on American technological ingenuity, which de-emphasises scientific uncertainties in order to focus on green innovations and markets, is even flexible enough to accommodate climate change sceptics, including former President George W. Bush, who recently declared, 'I'm confident that with sensible and balanced policies from Washington, American innovators and entrepreneurs will pioneer a new generation of technology that improves our environment, strengthens our economy, and continues to amaze the world' (White House 2008).

Conclusion: don't waste a good crisis

As Fischer (2003, p. 13) argues, 'basic to the politics of policymaking ... must be an understanding of the discursive struggle to create and control systems of shared social meanings'. This analysis has evaluated the struggles to create a shared system of meaning that could transform a circular debate between 'climate sceptics' and 'climate believers' into a broadly acceptable programme for large-scale US policy action on climate change. It argues that the economic opportunity frame for US climate change policy deployed by the Apollo Alliance represents a positive and flexible discursive strategy for proponents of large-scale climate change mitigation in the USA, as it provides a positive rationale for economic transformation towards a clean energy economy and integrates more disparate interests than either the risk and information frame or the securitisation frame.

Despite the comparative domestic strengths of the economic opportunity frame, however, the dynamics of climate change policy cannot be separated from the larger globalisation meta-discourse within which all nation states contend. Fogel (2007, p. 101) refers to the process of 'discourse institutionalisation' via which 'discourses solidify into institutions through policies, organisational practices, or dominant ways of reasoning'. This study adapts this concept to frames, as these must also eventually stabilise in order to shape long-term public policy trajectories. The stabilisation process, in turn, is embedded in a web of global economic and political structures that exert external constraints on the national priorities of even a country as powerful as the USA. For US climate policy the key issue at present is whether any long-term commitment to action against global warming can survive the global financial crisis precipitated by the collapse of the US mortgage market and exacerbated by the ensuing chaos in the international financial system. Predictions about the potential depth and length of the crisis are not germane to this analysis. The crucial point is how quickly structural dynamics can change within a global and tightly linked financial system and how these changes, in turn, affect the construction of policy problems and throw priorities, policies and worldviews back into contention.

For example the Apollo Alliance, whose institutional and political power was substantially augmented by President Obama's electoral victory in November 2008, wants to embed its ambitious industrial transformation programme within the American Recovery and Reinvestment Act of 2009. The Recovery Act, described by the Department of Energy as 'an extraordinary response to a crisis unlike any since the Great Depression', authorises $787 billion to such priorities as economic recovery, creating and preserving jobs, and stabilising state and local government budgets. It includes $32.8 billion for energy efficiency, investment and research. Trying not to waste a good crisis, the Apollo Alliance links its previous emphasis on green transformation with the financial and innovative opportunities opened up by the federal response to the global economic crisis. Yet in April 2009 the US unemployment rate reached 8.5%, a level not seen since 1983. The Recovery Act could propel new growth in 'green jobs', defined by the Center on Wisconsin Strategy (2008, p. 2) as 'family supporting, middle-skill jobs in the primary sectors of a clean energy economy'. However, it is also possible that the green transformation/opportunity frame could be derailed by a prolonged economic crisis that recasts political and ideological divisions and revitalises the post-Cold War salience of class-based economic analysis and action. Against the backdrop of the financial crisis, the interests of labour, environmentalists, national security actors and corporate stakeholders could prove too diverse to be contained within a climate change initiative. Consequently, as Steensland (2008, p. 1030) puts it, the next stage in the evolution of frame analysis as a method is to devote more attention to analysing how and why frames enter and exit the arena of public discourse.

References

Apollo Alliance, 2008. *The New Apollo Program: clean energy, good jobs*. San Francisco: The Apollo Alliance.

Barker, C. and Willis, P., 2003. *Cultural studies: theory and practice*. 2nd ed. London: Sage Publications.

Barnett, J., 2001. *Security and climate change*. Working Paper 7 (October). Norwich: Tyndall Centre for Climate Change Research.

Bird, E., 1987. The social construction of nature: theoretical approaches to the history of environmental problems. *Environmental Review*, 11 (4), 255–264.

Brown, O. and Crawford, A., 2008. Climate change: a new threat to stability in West Africa? Evidence from Ghana and Burkina Faso. *African Security Review*, 17 (3), 39–57.

Bueren, E.M., Klijn, E.H., and Koppenjan, J.F.M., 2003. Dealing with wicked problems in networks: analysing an environmental debate from a network perspective. *Journal of Public Administration Research and Theory*, 13 (2), 193–212.

Busby, J.W., 2007. *Climate change and national security: an agenda for action*. CSR No.32. New York: Council on Foreign Relations.

Buzan, B., Waever, O., and de Wilde, J., 1998. *Security: a new framework for analysis*. Boulder, CO and London: Lynne Rienner Publishers.

Center on Wisconsin Strategy, The Workforce Alliance, and the Apollo Alliance, 2008. *Greener pathways: jobs and workforce development in the clean energy economy*. Madison, WI: Center on Wisconsin Strategy.

CNA Corporation, 2007. *National security and the threat of climate change* [online]. Alexandria, VA: The CNA Corporation. Available from: http://securityandclimate. can.org [Accessed 26 October 2008].

Creed, W.E.D., Langstraat, J.A., and Scully, M.A., 2002. A picture of the frame: frame analysis as technique and as politics. *Organizational research methods*, 5 (1), 34–55.

Danto, A.C., 1965. *Nietzche as philosopher*. New York: Macmillan.

Depledge, J., 2006. The opposite of learning: ossification in the climate change regime. *Global Environmental Politics*, 6 (1), 1–22.

Demeritt, D., 2002. What is the 'social construction of nature'? A typology and sympathetic critique. *Progress in Human Geography*, 26 (6), 767–790.

Fischer, F., 2003. *Reframing public policy: discursive politics and deliberative processes*. Oxford: Oxford University Press.

Fogel, C., 2007. Constructing progressive climate change norms: the US in the early 2000s. *In*: M.E. Pettenger, ed. *The social construction of climate change: power, knowledge, norms, discourses*. Burlington, VT: Ashgate Publishing Company, 99–122.

Goffman, E., 1974. *Frame analysis*. New York: Harper and Row Books.

Gore, A., 2007. *Nobel lecture, 10 December 2007* [online]. The Nobel Foundation. Available from: http://nobelprize.org [Accessed 2 November 2008].

Hajer, M., 1993. Discourse coalitions and the institutionalisation of practice: the case of acid rain in Britain. *In*: F. Fischer and J. Forester, eds. *The argumentative turn in policy analysis and planning*. Durham, NC and London: Duke University Press, 43–76.

Inglis, B. and Laffer, A.B., 2008. Op-ed: an emissions plan conservatives could warm up to. *The New York Times*, 28 December.

Ingstad, B. and Whyte, S., 1995. *Disability and culture*. Berkeley, CA: University of California Press.

Inslee, J. (Representative), 2005. *The New Apollo Energy Act* [online]. United States House of Representatives. Available from: http://www.house.gov/inslee [Accessed 30 October 2008].

Johnson, K., 2006. Framing AIDS mobilization and human rights in post-apartheid South Africa. *Perspectives on Politics*, 4 (4), 663–670.

Kennedy, J.F., 1961. Special message to Congress on urgent national needs, 25 May [online]. Available from: http://www.jfklibrary.org/Historical+Resources/Archives/ Reference+Desk/Speeches/JFK/003POF03NationalNeeds05251961.htm [Accessed 7 August 2009].

Lynas, M., 2008. Climate change catastrophe by degrees [online]. *The Guardian*, 7 August. Available from: http://www.guardian.co.uk/commentisfree/2008/aug/07/ carbonemissions.climatechange [Accessed 9 January 2009].

Mathews, J.T., 1989. Redefining security. *Foreign Affairs*, 68 (Spring), 162–177.

Norris, R., 2008. *Lessons of the Manhattan Project* [online]. *In: A presentation to the National Academies' Committee on Science, Engineering and Public Policy*, 5 September. Available from: http://docs.nrdc.org/nuclear/files/nuc_08100901A.pdf [Accessed 9 November 2008].

Nygren, A. and Rikoon, S., 2008. Political ecology revisited: integration of politics and ecology does matter. *Society and Natural Resources*, 21, 767–782.

Phillips, N. and Hardy, C., 2002. *Discourse analysis: investigating the processes of social construction*. London: Sage Publications.

Pickerill, J., 2008. From wilderness to WildCountry: the power of language in environmental policy campaigns in Australia. *Environmental politics*, 17 (1), 95– 104.

Reber, B.H. and Berger, B.K., 2005. Framing analysis of activist rhetoric: how the Sierra Club succeeds or fails at creating salient messages. *Public Relations Review*, 31, 185–195.

Saarikoski, H., 2006. When frames conflict: policy dialogue on waste. *Environment and Planning C: Government and Policy*, 24, 615–630.

Schön, D.A. and Rein, M., 1994. *Frame reflection: toward the resolution of intractable policy controversies*. New York: Basic Books.

SCOPE, 2008. Green jobs: The Los Angeles Apollo Alliance [online]. Available from: http://www.scopela.org [Accessed 11 December 2008].

Shmueli, D., Elliott, M., and Kaufman, S., 2006. Frame changes and the management of intractable conflicts. Available from: http://urban.csuohio.edu/~sanda/papers/ intractable.pdf [Accessed 4 January 2009].

Steensland, B., 2008. Why do policy frames change? Actor-idea coevolution in debates over welfare reform. *Social Forces*, 86 (3), 1027–1053.

Ullman, R., 1983. Redefining security. *International Security*, 8 (1), 129–153.

United Nations, 2000. *UN Security Council Resolution 1308(2000) on the responsibility of the Security Council in the maintenance of international peace and security: HIV/ AIDS and international peace-keeping operations* [online]. Available from: http:// uniformservices.unaids.org [Accessed 30 October 2008].

United Nations, 2007. *Security Council holds first-ever debate on impact of climate change on peace, security, hearing over 50 speaker*, 17 April [online]. Available from: http://www.un.org/News/Press/docs/2007/sc9000.doc.htm [Accessed 7 August 2009].

United Nations, 2008. *Unless action is taken on a war-footing to address climate change, world will miss Millennium development goals* [online]. United Nations. Available from: http://www.un.org/News/Press/docs/2008/dsgsm404.htm [Accessed 30 October 2008].

Waever, O., 1999. Securitizing sectors? Reply to Eriksson. *Cooperation and Conflict*, 34 (3), 334–340.

Walsh, B., 2008. Does global warming compromise national security? *Time*, 16 April [online]. Available from: http://www.time.com/time/specials/2007/article/0,28804, 1730759,00.html [Accessed 30 October 2008].

Watson, H., 2005. *US climate change policy overview*. Presented at the Seminar of Government Experts, 16 May, Bonn, Germany.

White House, 2001a. Text of a letter from the President to Senators Hagel, Helms, Craig and Roberts, 13 March [online]. Available from: http://www.whitehouse.gov/news/releases/2001/03/20010314.html [Accessed 9 January 2009].

White House, 2001b. President Bush discusses global climate change, 11 June [online]. Available from: http://www.whitehouse.gov/news/releases/2001/06/20010611-2.html [Accessed 31 March 2009].

White House, 2008. President Bush discusses climate change [online]. Available from: http://www.whitehouse.gov/news/releases/2008/04/20080416-6.html [Accessed 7 January 2009].

Wildavsky, A., 1979. *Speaking truth to power: the art and craft of policy analysis*. New York: Little, Brown and Co.

Index

Page numbers in *Italics* represent tables.
Page numbers in **Bold** represent figures.

For Product Safety Concerns and Information please contact our EU
representative GPSR@taylorandfrancis.com
Taylor & Francis Verlag GmbH, Kaufingerstraße 24, 80331 München, Germany